The Microbiomes of Humans, Animals, Plants, and the Environment

Volume 7

Series Editor
Andrés Moya
I2Sysbio
University of València
Paterna, Spain

This series covers microbiome topics from all natural habitats. Microbiome research is a vibrant field of science that offers a new perspective on Microbiology with a more comprehensive view on different microorganisms (microbiota) living and working together as a community (microbiome). Even though microbial communities in the environment have long been examined, this scientific movement also follows the increasing interest in microbiomes from humans, animals and plants. First and foremost, microbiome research tries to unravel how individual species within the community influence and communicate with each other. Additionally, scientists explore the delicate relationship between a microbiome and its habitat, as small changes in either, can have a profound impact on the other. With individual research volumes, this series reflects the vast diversity of Microbiomes and highlights the impact of this field in Microbiology.

Christian Rinke • Mária Džunková
Editors

Single-Cell Omics in Microbiome Research

Exploring Microbial Diversity Across Ecosystems

Springer

Editors
Christian Rinke
Department of Microbiology
The University of Innsbruck
Innsbruck, Austria

Mária Džunková
Institute for Integrative Systems Biology
(I2SysBio)
University of Valencia Science Park
Paterna, Valencia, Spain

ISSN 2662-611X ISSN 2662-6128 (electronic)
The Microbiomes of Humans, Animals, Plants, and the Environment
ISBN 978-3-032-07526-0 ISBN 978-3-032-07527-7 (eBook)
https://doi.org/10.1007/978-3-032-07527-7

This Springer imprint is published by the registered company Springer Nature Switzerland AG
The registered company address is: Gewerbestrasse 11, 6330 Cham, Switzerland

Contents

A Hitchhikers Guide to Microbial Single-Cell Omics

1

Christian Rinke

Abstract

Microbial single-cell omics has revolutionized the study of uncultured microorganisms, offering unparalleled insights into their genetic composition, functional potential, and ecological roles. This guide highlights the key advancements in single-cell genomics, transcriptomics, proteomics, and metabolomics as featured in this book. It explores innovative techniques that enhance genome recovery, boost throughput, and lower costs. Additionally, new activity-based methods allow precise targeting of specific microbial populations, while the integration of multi-omics workflows represents a major leap forward. These breakthroughs are set to transform our understanding of microbial diversity and function, with profound implications for ecology, biomedicine, and biotechnology.

Keywords

Microbial single-cell omics · Fluorescence-activated cell sorting (FACS) · Multiple displacement amplification (MDA) · Hybridization Chain Reaction (HCR)-FISH · Multi-omics integration

Microorganisms are essential to maintaining Earth's biodiversity and ecosystems. Yet, the vast majority of microbes continue to evade traditional culturing methods. This challenge can now be addressed using cultivation-independent approaches, such as microbial single-cell omics, a suite of methods that provide powerful

C. Rinke (✉)
Department of Microbiology, The University of Innsbruck, Innsbruck, Austria
e-mail: christian.rinke@uibk.ac.at

C. Rinke, M. Džunková (eds.), *Single-Cell Omics in Microbiome Research*, The Microbiomes of Humans, Animals, Plants, and the Environment 7,
https://doi.org/10.1007/978-3-032-07527-7_1

insights into the genetic composition and functional potential of individual microbial cells. Since the first application of single-cell genomics to bacteria in 2005 (Raghunathan et al. 2005), it has remained a rapidly evolving approach, playing a pivotal role in diverse research areas, including the discovery and characterization of new bacterial and archaeal lineages (Kamke et al. 2014; Nobu et al. 2016; Rinke et al. 2013), expanding our understanding of the fungal tree of life (Ahrendt et al. 2018), resolving genomic variations between bacterial strains (Kashtan et al. 2014), and revealing virus–host interactions (Džunková et al. 2019; Stepanauskas et al. 2017), to name but a few.

The field of microbial single-cell genomics has made significant strides since the pioneering studies that guided our understanding of genome amplification (Lasken and Stockwell 2007), contamination prevention (Woyke et al. 2011), volume minimization (Marcy et al. 2007a), microfluidic applications (Marcy et al. 2007b), and fluorescence-activated cell sorting (Rinke et al. 2014). Over the years, multiple displacement amplification (MDA) has remained the method of choice for microbial single-cell genomics, outperforming all alternatives. MDA innovations such as a thermostable mutant of the phi29 polymerase (Stepanauskas et al. 2017) and the introduction of exonuclease-resistant terminators (Gonzalez-Pena et al. 2021; Bowers et al. 2024) have further improved genome quality and completeness.

In addition to technological advances that push the boundaries of genome recovery, throughput maximization, and cost reduction, the broader field of single-cell omics is also expanding towards encompassing the full range of omics approaches. These include transcriptomics, proteomics, and metabolomics, enabling the analysis of transcriptional activity, protein content, and metabolite composition of single bacterial and archaeal cells. The book titled "Single-Cell Omics in Microbiome Research" is a testament to the recent progress, and offers a focused outlook on expected developments over the next decade, particularly in light of breakthroughs in high-throughput multiplexing and the transformative impact of the AI revolution on software development.

But let's start with the basics. Any venture into working with single microbial cells is considerably accelerated by following an expert protocol that outlines *best-practice laboratory workflows*. Goudeau and Malstrom (DOE-Joint Genome Institute, USA) provide precisely this foundation for microbial single-cell genomics in the second chapter of this book (see Chap. 2). Key topics include prokaryotic sample preservation and preparation, fluorescence-activated cell sorting (FACS), and multiple displacement amplification (MDA), along with common challenges and effective strategies to address them. The authors share their expertise in identifying microbial populations and sorting individual cells on FACS instruments, while also detailing how UV treatments optimize reaction kinetics during MDA reactions.

Once a single-cell genomics pipeline is operational, the next goals are typically to increase throughput while simultaneously reducing per-cell reagent costs. Microfluidic workflows, particularly droplet-based approaches, offer considerable promise to achieve both of these tasks. However, most procedures for prokaryotes have remained lengthy and rely on custom devices that require expert knowledge. In the third chapter of this book (see Chap. 3), Ortiz-Maiques and Džunková (University of Valencia and the Spanish National Research Council, Spain) describe a recently introduced *microfluidic single-cell encapsulation* and amplification method that has the potential to transform the status quo. The approach incorporates two commercially available methods. First, a droplet-based microfluidics workflow, based on semi-permeable capsules (SPCs), is applied to create thousands of encapsulated single cells within one tube. Next, a recently developed whole genome amplification procedure, termed primary template-directed amplification (PTA), produces single amplified genomes (SAGs) with a more uniform genome coverage. Finally, the approach benefits from a split-and-pool barcoding process that allows ligation-based multiplexing of over 10,000 SAGs into a single library, resulting in a considerable reduction of library creation costs. A detailed protocol is provided for two workflows: the high-throughput Label—Sort—Encapsulate—Amplify (LSEA) procedure, employing whole genome amplification, split-and-pool barcoding, and long-read sequencing for genome recovery, and the Encapsulate—Amplify—Label—Sort (EALS) procedure to amplify genes of interest in a capsule-PCR process that uses fluorescently labelled primers.

Most single-cell genomics approaches target all microbial cells in a community, while allowing some degree of specificity based on cell size, DNA stain intensity, and granularity (Rinke et al. 2014). However, certain research aims require *targeted approaches* to recover microbes of interest. The combination of fluorescence in situ hybridization (FISH) with fluorescence-activated cell sorting (FACS) enables targeting uncultured microbes and is well suited to recover members of the rare biosphere. This approach, termed *FISH–FACS*, is detailed in chapter four (see Chap. 4), via a collaborative effort by the teams of Woyke (Department of Energy Joint Genome Institute, USA) and Fuchs (Max Planck Institute for Marine Microbiology, Germany). The authors introduce a hybridization chain reaction (HCR)–FISH protocol that provides a strong FISH signal even for small cells with low ribosome content and avoids harsh cell fixation to support genome amplification and downstream sequencing. This approach relies on the presence of 16S rRNA sequence data, which are needed to design FISH probes for target lineages in the sample. HCR–FISH is therefore not suited to recover novel, divergent lineages, but it is an excellent tool to study taxa that have only sparse genomic representation. The chapter concludes with a detailed experimental design section, including cell fixation, labeling with a target probe and initiator sequence, followed by a chain reaction to boost the fluorescent signal, and cell sorting via FACS. A detailed list of materials and step-by-step instructions creates a strong urge to try FISH-FACS as soon as possible.

Another promising method to target single cells of interest is the *activity-based separation* of microbial communities. In chapter five (see Chap. 5), Reichart (Pacific Northwest National Laboratory, USA) reviews two of the latest click chemistry approaches and their applications for activity-based separation by FACS. *BioOrthogonal Non-Canonical Amino acid Tagging (BONCAT)* uses non-natural amino acids to label and track newly synthesized proteins. Benefits include low costs, and the ability to track any translational activity regardless of the substrate; however, high reagent concentrations and long incubation times can adversely impact cell growth. Applications include recovering active microbial populations in soil, screening for metabolic activity of hot spring sediment slurries, assessing the degradation of polysaccharides in deep-sea sediments and deciphering active microbes in the sputum of Cystic fibrosis patients. In contrast, *Activity-Based Protein Profiling (ABPP)* is substrate-specific since it relies on molecular probes designed to mimic enzymatic substrates, including organic matter, metabolite products, or even virulence factors. A considerable advantage is that ABPP probes can be “pre-clicked” with a fluorophore, avoiding exposure of the microbial community to click chemistry reagents, facilitating the recovery of viable cells. Applications for microbial cells are still sparse, but examples, such as screening soil microbiomes for glycosidase activity and subsequent FACS separation of identified active cells, are a promising start.

In chapter number six (see Chap. 6), Takeyama, Hosokawa, Ando, and Nishikawa (Waseda University, Japan) review *droplet-based single-cell genomics*. The authors initially focus on the recovery of small viral genomes through amplification within gel beads, and on targeting bacterial genomes of interest through a PCR step, followed by the detection with fluorescence probes. Next, the chapter moves on to recent methodological advances and applications of *microbial single-cell transcriptomics*. Benefits and limitations of a wide range of approaches are covered, such as split-pool barcoding for high-throughput analysis, and droplet-based methods using barcoded beads to profile large bacterial populations. Applications of these methods have revealed functional heterogeneity within microbial communities, have helped to explore subpopulations that are able to withstand antibiotic treatment, and have aided in the characterization of non-coding RNAs. The final focus point in this chapter is *microbial single-cell metabolomics* and touches on the challenges of integrating microfluidic devices with mass spectrometry analysis and on overcoming current detection limits. Advantages of confocal Raman microspectroscopy, providing nondestructive, high-resolution imaging of metabolites, are discussed, as well as the incorporation of alkyne tags into Raman analysis and the combination of Raman with stable isotope probing (SIP), to enable semi-quantitative assessments of metabolic activity at the single-cell level. Recent computational advances now enable multivariate curve resolution (MCR) analysis to separate overlapping spectral signatures and to generate quantitative spatial distribution maps of specific metabolites within single cells. The authors conclude that the integration of single-cell omics technologies with AI-driven analytics in the near future will unlock new possibilities in microbial ecology, biomedicine, and biotechnology.

Developing multi-omics workflows, such as creating genomic and transcriptomic data from a single bacterial or archaeal cell, is especially important for the study of uncultured microbial lineages that lack genomic references to map transcriptomic reads. Magold and Kaster (Karlsruhe Institute of Technology, Germany) address this challenge in chapter seven (see Chap. 7), by reviewing single-cell *multi-omics techniques* and exploring their applications to *uncultured prokaryotic cells*. A major obstacle remains the task of simultaneously recovering cDNA (reverse transcribed RNA) and genomic DNA from a single prokaryotic cell. The authors discuss possible workflows to achieve this goal by considering, for example, the recent development of transposase for DNA barcoding, reverse transcription based on primers targeting internal regions of RNA transcripts, and a CRISPR-Cas9-based rRNA depletion step. Approaches combining genomics and proteomics of single prokaryotic cells are equally challenging, although DNA amplification followed by protein-targeting antibodies might be a way forward to recover at least known proteins from microbial dark matter taxa. Finally, the authors pull out all the stops and discuss ways to combine prokaryotic single-cell transcriptomics and proteomics, a daunting task given that currently both individual steps represent a considerable challenge. However, suggestions such as splitting the sample after cell lysis and processing the RNA and protein fractions separately, and recently published methods combining single-cell proteomics and metabolomics in eukaryotes, based on mass spectrometry, are inspiring optimism.

The final chapter of this book (see Chap. 8), authored by Mora-Quilis and Domingo-Calap (University of Valencia, Spain), explores the impact of bacterial heterogeneity on *phage–host interactions*. The authors employ single-cell sorting to isolate individual bacterial cells and examine their phenotypes, i.e., phage resistance, under a variety of conditions. They highlight that while bacteria can evade viral predation by altering cell surface proteins, this strategy may inadvertently render them vulnerable to previously non-infective phages or even antibiotics. Importantly, these structural changes are not solely driven by genetic mutations but also by shifts in gene expression, underscoring the need to isolate single cells from heterogeneous populations to study these transient phenotypes. The chapter provides a comprehensive protocol, detailing the materials and methods required to prepare phage and bacterial suspensions, perform bacterial single-cell sorting using FACS, and conduct subsequent incubation and counting procedures.

In *summary*, single-cell omics has advanced considerably over the past decades, and recent progress in microbial single-cell transcriptomics, proteomics, and metabolomics, combined with the development of AI-based tools, suggests that the coming decade holds the potential for even more transformative breakthroughs.

References

Ahrendt SR, Alisha Quandt C, Ciobanu D et al (2018) Leveraging single-cell genomics to expand the fungal tree of life. Nat Microbiol 3(12):12. https://doi.org/10.1038/s41564-018-0261-0

Bowers RM, Gonzalez-Pena V, Wardhani K et al (2024) scMicrobe PTA: near complete genomes from single bacterial cells. ISME Commun 4(1):ycae085. https://doi.org/10.1093/ismeco/ycae085

Džunková M, Low SJ, Daly JN, Deng L, Rinke C, Hugenholtz P (2019) Defining the human gut host–phage network through single-cell viral tagging. Nat Microbiol 4(12):2192–2203

Gonzalez-Pena V, Natarajan S, Xia Y et al (2021) Accurate genomic variant detection in single cells with primary template-directed amplification. Proc Natl Acad Sci 118(24):e2024176118. https://doi.org/10.1073/pnas.2024176118

Kamke J, Rinke C, Schwientek P et al (2014) The candidate phylum poribacteria by single-cell genomics: new insights into phylogeny, cell-compartmentation, eukaryote-like repeat proteins, and other genomic features. PloS One 9(1):e87353

Kashtan N, Roggensack SE, Rodrigue S et al (2014) Single-cell genomics reveals hundreds of coexisting subpopulations in wild prochlorococcus. *Science* 344(6182):416–420. https://doi.org/10.1126/science.1248575

Lasken R, Stockwell T (2007) Mechanism of chimera formation during the multiple displacement amplification reaction. BMC Biotechnol 7(1):19. https://doi.org/10.1186/1472-6750-7-19

Marcy Y, Ishoey T, Lasken RS et al (2007a) Nanoliter reactors improve multiple displacement amplification of genomes from single cells. PLoS Genet 3(9):e155. https://doi.org/10.1371/journal.pgen.0030155

Marcy Y, Ouverney C, Bik EM et al (2007b) Dissecting biological "dark matter" with single-cell genetic analysis of rare and uncultivated TM7 microbes from the human mouth. Proc Natl Acad Sci 104(29):11889–11894. https://doi.org/10.1073/pnas.0704662104

Nobu MK, Dodsworth JA, Murugapiran SK et al (2016) Phylogeny and physiology of candidate phylum "atribacteria" (OP9/JS1) inferred from cultivation-independent genomics. ISME J. https://doi.org/10.1038/ismej.2015.97

Raghunathan A, Ferguson HR, Bornarth CJ, Song W, Driscoll M, Lasken RS (2005) Genomic DNA amplification from a single bacterium. Appl Environ Microbiol 71(6):3342–3347. https://doi.org/10.1128/AEM.71.6.3342-3347.2005

Rinke C, Schwientek P, Sczyrba A et al (2013) Insights into the phylogeny and coding potential of microbial dark matter. Nature 499(7459):431–437

Rinke C, Lee J, Nath N et al (2014) Obtaining genomes from uncultivated environmental microorganisms using FACS–based single-cell genomics. Nat Protoc 9(5):1038–1048

Stepanauskas R, Fergusson EA, Brown J et al (2017) Improved genome recovery and integrated cell-size analyses of individual uncultured microbial cells and viral particles. Nat Commun 8(1):84. https://doi.org/10.1038/s41467-017-00128-z

Woyke T, Sczyrba A, Lee J et al (2011) Decontamination of MDA reagents for single cell whole genome amplification. PloS One 6(10):e26161. https://doi.org/10.1371/journal.pone.0026161

2 Overview of Laboratory Workflows for Microbial Single-Cell Genomics

Danielle Goudeau and Rex R. Malmstrom

Abstract

Single-cell genomics is a powerful tool for recovering the genome sequences of the uncultivated species comprising complex microbial communities. The laboratory procedure typically involves sorting individual microbial cells into reaction wells before performing whole-genome amplification. Here, we provide an overview of laboratory steps such as sample preservation, single-cell sorting, and performing whole-genome amplification reactions. We also discuss common challenges and provide suggestions for dealing with problems such as DNA contamination that can derail single-cell genomics studies.

Keywords

Single-cell genomics · Multiple Displacement Amplification · Fluorescence Activated Cell Sorting · SAG

2.1 Background

Microorganisms are critical drivers of energy and nutrient flow through nearly all biological systems (Falkowski et al. 2008). Their genomes also encode a tremendous diversity of metabolic capabilities that can be harnessed to enhance plant growth (Abhilash et al. 2016), increase carbon sequestration (Tao et al. 2023), and manufacture valuable bioproducts (Keasling 2010). However, most species are not easily cultured in the laboratory (Rappé and Giovannoni 2003), so

D. Goudeau · R. R. Malmstrom (✉)
DOE Joint Genome Institute, Lawrence Berkeley National Laboratory, Berkeley, CA, USA
e-mail: rrmalmstrom@lbl.gov

C. Rinke, M. Džunková (eds.), *Single-Cell Omics in Microbiome Research*, The Microbiomes of Humans, Animals, Plants, and the Environment 7,
https://doi.org/10.1007/978-3-032-07527-7_2

culture-independent methods such as metagenomics and single-cell genomics are used to explore their genetic makeup. In contrast to metagenome-assembled genomes (MAGs), which are composites of highly similar genomes (Sangwan et al. 2016), all DNA sequences in a single-cell amplified genome (SAG) originate from the same cell. Linking genetic variations across a genome and establishing connections between mobile elements, such as viruses, and host genomes is easier when examining DNA derived from individual cells (Labonté et al. 2015; Roux et al. 2014; Jarett et al. 2020). Single-cell genomics has also shed light on uncultured microbial 'dark matter' inhabiting a wide variety of systems (Pachiadaki et al. 2017, 2019; Swan et al. 2011; Bowers et al. 2022; Rinke et al. 2013; Engel et al. 2014; Marcy et al. 2007a; Zheng et al. 2022). However, single-cell genomics methods can be difficult to adopt and scale (Rinke et al. 2014). This chapter provides an overview of the laboratory process and highlights important considerations for carrying out microbial single-cell genomics projects.

The laboratory process for single-cell has two fundamental steps, each with its own difficulties and technical considerations. The first involves physically separating individual cells into different compartments, such as microtiter plate wells, microdroplet emulsions, or microscopic hydrogel beads. This physical separation is followed by the second step, whole-genome amplification (WGA). The large quantities of DNA generated during WGA can be uniquely molecularly barcoded during sequencing library creation, thanks to the initial physical separation of single cells into individual reaction chambers. These barcodes identify DNA derived from the same cell after pooling hundreds of uniquely labeled single-cell libraries for large-scale DNA sequencing. DNA sequences sharing the same molecular barcode, i.e., sharing the same origin, can be assembled de novo for genome reconstruction, mapped to reference genomes to identify genomic variation, mined for novel metabolic pathways, and subjected to any number of other analyses.

Various methods such as micromanipulation (Woyke et al. 2010), serial dilution (Zhang et al. 2006), and optofluidics (Blainey et al. 2011) have all been used to isolate individual cells, but fluorescence-activated cell sorting (FACS) is currently the predominant method in single-cell genomic studies (Kaster and Sobol 2020). Traditional FACS instruments sort individual cells directly into 96-well and 384-well plates, where subsequent lysis and WGA reactions are carried out. This approach makes sorting thousands of individual cells in an afternoon possible, thus enabling high-throughput single-cell genomics pipelines (Rinke et al. 2014; Rodrigue et al. 2009; He et al. 2016). A variation of this approach entails encasing individual cells inside hydrogel beads or semi-permeable capsules before lysis and WGA (Nishikawa et al. 2022; Leonaviciene et al. 2020; Chijiiwa et al. 2020). Lysis and WGA reagents diffuse in and out of gel beads while amplified DNA remains trapped inside. Individual gel beads containing amplified DNA are then deposited into microwell plates using FACS before creating sequencing libraries containing unique molecular barcodes (Nishikawa et al. 2022; Chijiiwa et al. 2020). Some recent studies used sophisticated microfluidics procedures to introduce unique molecular barcodes directly into microdroplet emulsions containing single cells, thus bypassing the need for FACS altogether (Zheng et al. 2022; Aoki et al. 2021).

These FACS-free solutions have great potential, but currently, genome quality is highly variable and the sample throughput is limited. The remainder of this chapter will focus on the traditional flow sorting of single cells into microwell plates, followed by genome amplification and sequencing.

In single-cell genomics studies, the primary goal of FACS is to identify individual cells and sort them one at a time into their own wells of a microtiter plate. A FACS instrument focuses a liquid cell suspension into a narrow stream so that cells align one after another before passing through a detector illuminated with lasers. The scattering of laser light provides information about the size and composition of each cell and can detect doublets or aggregates of cells (Shapiro 2003a).

Cells are typically identified by a combination of scatter and fluorescence signals and general DNA stains that distinguish cells from debris in environmental samples. Most single-cell studies utilize generic DNA staining techniques to identify and randomly sort cells (Rinke et al. 2014; Stepanauskas et al. 2017). Alternatively, specific stains and labeling strategies can be used to select subsets of cells based on phylogenetic or phenotypic characteristics. For example, pigmented cells with distinct autofluorescent signals can be specifically recovered (Kashtan et al. 2014; Wu et al. 2021; Nakayama et al. 2019; Birzu et al. 2023), and for non-pigment cells, fluorescence in situ hybridization (FISH) probes can identify cells belonging to specific taxonomic groups (Grieb et al. 2020; Yilmaz et al. 2010; Podar et al. 2007). Functional stains and strategies like LysoTracker Green target cells with acidic food vacuoles, while RedoxSensor Green and BioOrthogonal NonCanonical Amino Acid Tagging (BONCAT) identify respiring cells or cells actively synthesizing protein, respectively (Munson-McGee et al. 2022; Couradeau et al. 2019; Hatzenpichler and Connon 2016; Needham et al. 2019). Later chapters discuss these powerful targeted cell selection strategies in greater detail.

After flow sorting individual cells, genome amplification is necessary to generate sufficient quantities of DNA for single-cell sequencing (Lasken 2007). Most bacterial cells contain just a few fg's of DNA (Lasken 2007), but sequencing library construction protocols typically require orders of magnitude more DNA, thus the need for WGA. Multiple displacement amplification (MDA) is the most common WGA method, and it uses a mixture of random hexamers to prime DNA amplification with phi29 polymerase to generate billions of genome copies (Dean et al. 2001). Phi29 is a highly processive, strand-displacing polymerase that can produce µg's of DNA from fg inputs (Lasken 2007; Dean et al. 2001; Raghunathan et al. 2005). Unfortunately, MDA does not amplify genomes uniformly, leading to wildly uneven coverage in sequencing results (Zhang et al. 2006; Rodrigue et al. 2009; Raghunathan et al. 2005; Marcy et al. 2007b). This uneven coverage hampers de novo assembly such that most single-cell genomes are $\leqq$ 40% complete (Rinke et al. 2013; Clingenpeel et al. 2014a). In some instances, using variations on amplification chemistry (e.g., LIANTI (Chen et al. 2017), MALBAC (de Bourcy et al. 2014), and WGAX (Stepanauskas et al. 2017)) and sample processing strategies (e.g., microcompartmentalization in emulsion droplets or hydrogel beads (Zheng et al. 2022; Nishikawa et al. 2022; Aoki et al. 2021; Lan et al. 2024; Wang et al. 2023; Abate et al. 2023)) have improved genome recovery, although none of these approaches appears to be a panacea. Interestingly, a quasi-linear amplification strategy known as primary template amplification (PTA) dramatically improves genome recovery from single human and bacterial cells by simply adding alpha-thiol ddNTPs to standard MDA reactions (Bowers et al. 2024; Gonzalez-Pena et al. 2021), but this new variation on MDA has yet to be applied to single-cell microbiome research at scale. Traditional MDA is still the predominant WGA method for microbial single-cell genomics, and it is the focus of the rest of this chapter.

The following sections highlight key steps in single-cell genomics laboratory workflows, including sample preservation and preparation, FACS instrument setup, sorting strategies, reagent decontamination, clean working procedures, and genome amplification. However, a unifying theme across these steps is "cleanliness". Single-cell genomics procedures are highly susceptible to contamination, and the few fg's of DNA from an individual bacterium can easily be swamped by exogenous DNA accidentally introduced into MDA reactions by researchers. The trace levels of DNA found on and in microwell plates, tubes, pipet tips, and molecular biology reagents–including the MDA reagents themselves–are also important sources of contamination that can compromise single-cell studies. Success depends on decontaminating instruments, labware, and reagents using UV radiation and bleach treatments, as well as avoiding the reintroduction of contamination by using clean handling procedures. The latter includes wearing clean disposable gloves, lab coats, arm/wrist covers, face masks, as well as working in PCR hoods or HEPA-filtered rooms whenever possible. Specific approaches to cleanliness are discussed in greater detail in the sections below.

2.2 Sample Preservation

While chemical fixation typically enhances fluorescence signals in flow cytometry assays, it harms genome amplification. Aldehyde-based fixatives usually inhibit MDA reactions entirely (Grieb et al. 2020; Clingenpeel et al. 2014b). Ethanol fixation also negatively affects MDA and genome recovery, although it can be employed for single-cell genomics studies when necessary (Clingenpeel et al. 2014b). For example, ethanol fixation benefits single-cell genomics studies utilizing fluorescent in situ hybridization (FISH) to target specific cell types (Grieb et al. 2020; Yilmaz et al. 2010; Podar et al. 2007). However, alcohol fixatives also distort autofluorescent signals from pigmented cells (Marie et al. 2014). Instead of chemical fixation, most single-cell studies preserve samples by flash freezing in liquid nitrogen after amending them with approximately 10% glycerol before storage at −80 °C before sorting (Rinke et al. 2013; Swan et al. 2013). Glycerol-based preservation works with various environmental samples for long term storage, including samples preserved for over 10 years (Birzu et al. 2023), although some cell loss occurs over time (Gasol and del Giorgio 2000).

2.3 Cleaning FACS Instrument

Strict cleaning and handling procedures are required to avoid introducing exogenous DNA contamination into single-cell WGA reactions (Lasken 2007; Woyke et al. 2011). Autoclaving of the sheath fluid is insufficient; removing all potential sources of contaminating DNA in the sheath fluid and fluidic system is necessary. Any DNA source (viable cells, dead cells, or free DNA) in the fluidic system can compromise single-cell studies. Running a solution of 10% household bleach (final conc. ~ 0.5% sodium hypochlorite) through the fluidic system for >1 h, followed by a rinse with several liters of sterile water, is an effective cleaning method (Rinke et al. 2014; Rodrigue et al. 2009). Some in-line filters are incompatible with bleach and should be removed from the fluidic system during bleach treatments to avoid filter degradation. However, in-line filters can also be a source of contamination if biofilms form inside them. Replacing filters regularly, e.g., before a new single-cell sorting experiment, is advantageous when filters cannot be treated during bleach cleaning. Single-use disposable sterile 0.2 μm filters can be used instead of original instrument filters on some machines. Unfortunately, bleach treatment is not always possible on some FACS instruments. In this case, the most stringent cleaning option approved by the manufacturer might be effective, but cleanliness must be verified with cell-free WGA reactions of empty sort droplets. See the section below on monitoring WGA reaction kinetics and reagent-only controls.

In addition to decontaminating the FACS fluidics plumbing, the sheath fluid must be free of contaminating DNA. Sheath fluid typically accounts for >90% of the sort droplet volume, so contaminated sheath fluid can overwhelm DNA from individual cells. Treatment with UV radiation is an effective method to remove DNA from sheath fluid. For example, sheath fluid can be placed in quartz flasks and stirred gently overnight under the UV lights of a PCR clean hood (Rinke et al. 2014; Rodrigue et al. 2009). Depending on the system, an empty sheath tank with its lid removed can be treated with UV simultaneously with the sheath fluid. Dedicating a sheath tank specifically for single-cell projects can help maintain clean sheath fluid. After a deep cleaning of the fluidics system, we found our BD Influx instrument could be used for four consecutive days with the same batch of UV-treated sheath without an increase in detectable contamination. However, this must be assessed on each system.

2.4 Preparing Samples for Sorting

Preparing samples for flow sorting typically entails creating a cell suspension of fluorescently labeled cells. There are many choices for resuspension solutions, and the optimal resuspension fluids are compatible with the sample in terms of osmotic potential and pH (Rinke et al. 2014). Filter-sterilized water from the same environment as the cells can aid in dilution, suspension, and disaggregation and may help preserve fragile cells. However, the farther the sample refractive index is from that of the sheath fluid, the more noise will interfere with small particle detection

(Shapiro 2003b). FACS instruments can process cell suspensions with concentrations that range over several orders of magnitude, and we target roughly 10^6 cells/ml whenever possible. Although FACS instruments can process higher cell concentrations, decreasing cell density can reduce the chance of two cells accidentally being co-sorted and increase sorting efficiency. Diluting samples in clean solutions, e.g., the UV-treated sheath fluid generated during instrument setup, also reduces free DNA present in the sample that may be co-sorted with the cell (Rodrigue et al. 2009).

With some environmental samples, disaggregation of cells and particles is necessary to generate a suspension of single cells. For example, in our hands, soil or rhizosphere samples often respond well to 1 minute of vortexing followed by 5 minutes of centrifugation at 500 rcf to pellet larger particles before collecting the supernatant (Schweitzer et al. 2022). Microbial mats or biofilms can be disaggregated by passing them ~20 times through a narrow gauge (e.g., 25) needle (Birzu et al. 2023). After disaggregation, passing the sample through a nylon filter whose pore size is <½ the FACS instrument tip diameter is necessary to remove large particles that can clog the instrument. Finally, the sample can be fluorescently labeled using generic DNA stains such as 1X SYBR or 500 nM SYTO dyes for at least 10 minutes in the dark before analysis. Fluorescent stains for other biomolecules (protein, lipids, etc.) and functional activities (respiration, protein translation, etc.) can also be applied, although the details are beyond the scope of this chapter.

2.5 Detecting and Selecting Cells for Sorting

Cells are typically identified by a combination of fluorescence and laser scatter signals; the former depends on the specific fluorescent dyes used, while the latter is related to cell size and granularity (Shapiro 2003c). Plotting scatter vs. fluorescence typically reveals a distinct population of particles whose scatter is consistent with small cells and whose DNA fluorescence signal is substantially higher than background instrument noise and fluorescent debris. A "gate" can be drawn around this population, instructing the instrument to sort only particles whose scatter and fluorescence fall within the gate (Fig. 2.1a, c). Detecting and gating bacteria is usually straightforward in cultures or aquatic samples. Still, it can be complicated in some samples with small, dimly fluorescent cells that generate a continuum of fluorescence that overlaps with noise and debris. In this case, testing different dyes and combinations of fluorescence and scatter signals may improve the signal-to-noise ratio. In addition, many flow cytometers struggle to detect bacteria by their scatter characteristics alone due to the cells' small size. Bacteria have similar, if not smaller, diameters than the wavelength of light used to detect them (e.g., 488 nm lasers) (Azam and Hodson 1977; National Research Council (US) Steering Group for the Workshop on Size Limits of Very Small Microorganisms 1999; Levin and Angert 2015), and it is easy to miss small cell populations. Some FACS instruments with special small particle detectors can measure scatter signals from environmental microbes, but users may have to rely on fluorescence signals alone to detect and

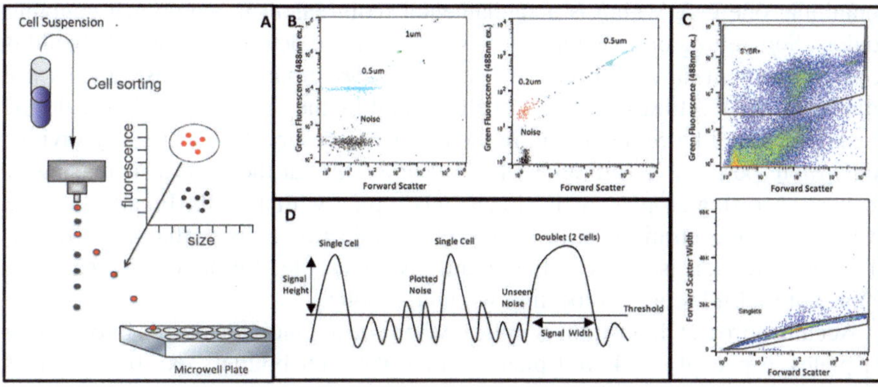

Fig. 2.1 (**a**) FACS overview. (**a**) Particles within a sample are analyzed based on scatter (an approximation for cell size) and fluorescence signals. Droplets containing specific particles can be sorted into microwell plates for downstream processing. (**b**) Examples of triggering detection on the green fluorescence channel for FACS without small particle detection (left) versus increased resolution when triggering on forward scatter with small particle detection (right). The 0.2 μm beads are indistinguishable from noise without the small particle detector in panel **b**. (**c**) Example of gating cells based on fluorescence using SYBR green DNA stain (top) and gating on single cells as a subset of these by comparing the height and width of the forward scatter signal (bottom). (**d**) Diagram illustrating how events/particles are defined on a FACS when the signal from a detector rises above and then falls below a designated threshold. A large signal width can signify ≥2 attached cells

identify cells (Fig. 2.1b). Analyzing small 0.2 μm calibration beads or a bacterial culture sample can help determine optimal detector settings.

Fluorescence and scatter signals are also helpful for distinguishing individual cells from multicell aggregates, a critical distinction in single-cell genomics studies. Doublets can be detected by plotting the DNA dye signal height by DNA dye signal width for the cell population in question (Fig. 2.1c, d). Most cells should form a diagonal line where the height and width are roughly proportional. A smaller percentage of events may deviate from this line by having a greater width than height. For populations of cells that are approximately the same size and shape, these deviating events can be classified as doublets or aggregates and excluded from the sort gates. For environmental samples with a mix of cell shapes, there is some chance of oddly shaped cells being misclassified and excluded from sorting, but this is balanced by an increased confidence in sorting only single cells.

2.6 Sorting Cells into Microwell Plates

FACS instruments usually have multiple sort options that optimize for different levels of purity (ensuring sort droplets contain only one target cell) versus yield (maximizing total recovery of target cells). Many instruments offer a stringent "single cell" setting designed to maximize the probability of sorting only droplets containing one cell, which is a fundamental requirement in single-cell genomics studies. Sort setting options typically use a "mask" to determine if a cell should be recovered or sent to waste because it cannot be sorted purely. For instruments with adjustable mask settings, the target cells should be the only particle in the droplet mask region and be centered within the droplet to ensure the sorting of individual cells. More permissive settings will speed up sorting at the cost of increasing the chances of sorting droplets with >1 cell and droplets with no cells. Finally, reducing overall cell concentrations or lowering sample injection rates so that the frequency of event detection is <1/50th the frequency of droplet formation decreases the chance of multiple cells co-locating within a droplet.

Accurate droplet deposition is critical when sorting individual cells into microwell plates, especially 384-well plates, where the well bottoms can be <2 mm in diameter. Cells accidentally deposited on well walls may not be covered by lysis and WGA reagents. Adjusting the microwell plates' X–Y positioning and the deflection angle of the sort stream may be necessary to deposit cells into the bottom of every well. Sorting roughly 100 droplets into each well allows operators to visualize sorting accuracy and to make fine-scale adjustments if droplets do not accumulate in the well bottoms. This approach provides greater insight than sorting droplets onto plates covered with film or foil, as droplets deflect at an angle and may still impact the well walls even though droplets appear centered on the film. Lastly, minimizing fanning—a phenomenon where droplets are deflected at different angles—is essential for accurate cell deposition. Most instruments have a camera where this can be visualized, though it can sometimes be hidden inside administrative settings. In our experience, fanning can often be reduced by decreasing sample injection rates, increasing the drop drive amplitude, and using a larger diameter sort tip when processing larger microbial cells.

Cells can be sorted "dry" into empty wells or "wet" into wells containing a buffer or lysis reagent. Either way, microwell plates should be rendered free of trace contaminating DNA before sorting, e.g., by exposing uncovered empty plates to UV radiation in a cross-linker for 10 minutes, adding buffer to wells if performing wet sorts, covering plates with transparent film, and treating with UV for another 10 minutes. Clean plates should remain covered until immediately before sorting and recovered with a fresh adhesive foil immediately after sorting to avoid introducing exogenous DNA contamination. Sorted plates can be sorted for months at −80 °C until ready to proceed with amplification.

Finally, many factors must be considered when designing a sort plate layout. Considering the potential impact of contamination in single-cell genomics, the number and location of negative control reactions are the most important factors. Reserving several wells (e.g., two columns) for negative controls is helpful. In this design, wells in one column receive an empty sort droplet, but not cells, to monitor the cleanliness of the sheath fluid and FACS instrument. In contrast, wells in another column remain empty to monitor WGA reagent cleanliness. Positive controls are another consideration. Adding tiny amounts of DNA to several wells is one option, but we prefer to sort 10–100 cells per well into one column (Rinke et al. 2014).

Lastly, the outer rows and columns can be accidentally contaminated during plate handling, and we recommend excluding these wells entirely.

2.7 Genome Amplification

After sorting, cells are ready for whole-genome amplification. Cell lysis and DNA release are necessary for whole-genome amplification to occur. Alkaline lysis with KOH or NaOH is the most common lysis method for MDA. Lysis protocols vary, but typically, cells are incubated for 5–30 min at pH > 13 before neutralization with Tris–HCl (Rinke et al. 2014; Raghunathan et al. 2005). Alkaline lysis is often carried out at ≦22 °C as colder temperatures can decrease DNA damage at high pH (Morono et al. 2014), although studies of some environments report greater success with warmer temperatures, such as 50–65 °C (He et al. 2016; Kvist et al. 2007). Before alkaline treatment, a series of freeze/thaw cycles is often commonly employed to enhance cell lysis (Kvist et al. 2007; Stepanauskas 2012). Additionally, lysozyme treatment of 5–50 units/μL for 15 minutes before alkaline lysis can increase the number of successful amplification reactions in some systems (Tsementzi et al. 2016). In some cases, treatment with proteinase K and nonionic detergents such as Tween may be helpful (Ciobanu et al. 2022). In all cases, enzymes and chemical lysis reagents are added sequentially to each sample well without DNA purification before genome amplification. Thus, lysis strategies must be chemically compatible with subsequent MDA reactions.

Following cell lysis, MDA reagents are added to the reaction wells to make many copies of the cell's genome. These reagents can be purchased as kits from different vendors or as individual components and assembled manually. Either way, great care must be taken to avoid contaminating the MDA reagents with exogenous DNA. MDA reactions use random hexamer primers that amplify contaminating DNA as easily as DNA from targeted microbial cells (Woyke et al. 2011; Ishoey et al. 2008). Since the bacterial cells contain just a few fg's of DNA, even trace amounts of contamination present a problem. Working cleanly reduces this risk. For example, using fresh disposable lab coats, gloves, and arm guards to avoid any exposed skin, paired with clean PCR hoods or HEPA-filtered room air, can reduce environmental contamination. UV contamination of all tubes, plates, and films will render them free of DNA.

Even with clean handling, DNA contamination still presents a problem since most MDA reagents also contain trace amounts of DNA that can interfere with microbial single-cell reactions (Woyke et al. 2010; Blainey and Quake 2011). For example, DNA sequences derived from endogenous reagent contaminants can be misassigned to target cell genomes. Fortunately, UV treatment of MDA reagents, including the Phi29 enzyme, can eliminate amplification in reagent-only, negative control reactions (Zhang et al. 2006; Rodrigue et al. 2009; Woyke et al. 2011). Aliquots of cell buffer, water, lysis, and stop reagents can all be UV-treated for moderate amounts of time (e.g., 30 minutes in a UV cross-linker) immediately before use with no ill effect. In contrast, dNTPs, hexamers, reaction buffers, and

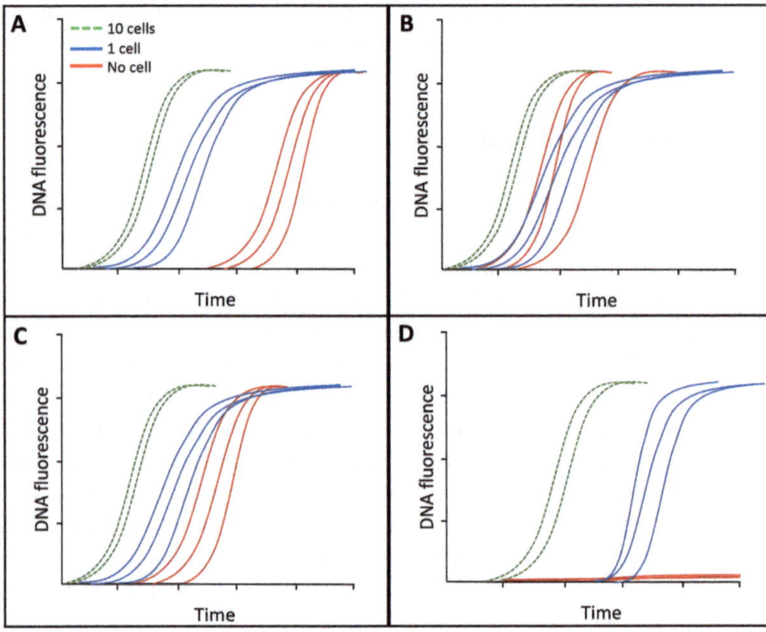

Fig. 2.2 MDA reaction kinetics and UV treatment to suppress amplification of reagent contamination. (**a**) Ideal spacing of amplification curves after successful UV treatment of MDA reagents produces a clear separation between positive controls (10 cells per reaction), single cells, and negative controls (reagent-only reactions). (**b**) Classic example of a dirty kit where negative controls amplify at comparable times to single-cell reactions. (**c**) Kinetics after some UV treatment, where negative controls amplify later, but still partially overlap with single-cell reactions. (**d**) Kinetics after too much UV treatment, where negative controls do not amplify at all, but positive controls and single cell reactions are substantially delayed

Phi29 polymerase are more fragile, and UV doses must be carefully titrated to avoid inhibiting MDA reactions altogether (Woyke et al. 2011). This can be accomplished by treating MDA reagents to different UV doses, monitoring reaction kinetics (Fig. 2.2), and selecting treatments that delay or suppress reagent-only reactions without substantially inhibiting the amplification in wells containing single cells. However, the exact dose must be determined empirically as the specific UV source and batch of reagents influence the cleaning efficacy. Unfortunately, working with contaminated commercial reagents is largely unavoidable, and it is necessary to physically eliminate contamination before MDA or informatically remove it after sequencing (Vollmers et al. 2022).

Monitoring amplification kinetics can reveal successful MDA reactions and possible contamination problems (Rinke et al. 2014; Zhang et al. 2006). Single-cell reactions reach completion at different times depending on factors such as the reagent composition and sample type. Within a plate, individual reactions will also reach completion at different times, possibly due to differences in lysis efficiency and genome accessibility among species (Rinke et al. 2014). In addition, reagent-only controls often produce large quantities of DNA due to contamination or an inherent propensity for generating primer-dimer-like artifacts (Lasken 2007). Unfortunately, endpoint measurements of total DNA concentrations cannot distinguish between wells containing clean, successful genome amplifications and wells containing contaminating DNA or amplification artifacts. However, with relatively clean reagents, these reagent-only controls often amplify later than single cells (Rinke et al. 2014; Woyke et al. 2011). Tracking reaction kinetics reveals which wells amplify before reagent-only controls, suggesting they contain clean and successful reactions, and which potentially problematic wells amplify at similar times to reagent-only controls. In addition, wells that amplify earlier tend to produce more complete genomes than wells that amplify later (Clingenpeel et al. 2014a, 2014b). Adding fluorescent dyes, e.g., 0.5 μM Syto, to MDA reactions enables real-time tracking of amplification kinetics using a qPCR instrument (Rinke et al. 2014). Fluorescence plate readers with temperature control are also an option since MDA is an isothermal reaction (Dean et al. 2001).

Reducing MDA reaction volumes can mitigate the impact of reagent contamination while lowering the overall costs of single-cell genomics projects. That is, the total amount of reagent-based contamination decreases when using fewer reagents per reaction, whereas the amount of target genome remains constant (Marcy et al. 2007b). Using fewer reagents per reaction also results in less cost per reaction. Reducing MDA reaction volumes entails proportionally decreasing all components of the buffer, lysis, neutralization, and MDA solutions while ensuring the final concentrations of each element remains unchanged (Rinke et al. 2014). The amount of buffer and lysis solution needed to cover the bottom of a reaction well and ensure contact with the sorted cell typically determines the final size of MDA reactions since lysis is the first step, and all other reagents are added proportionately. Different microwell plates require different volumes to cover well bottoms, so plate selection impacts the size of MDA reactions. Using liquid handlers with sub-microliter dispensing, MDA reactions of 2–3 μL are easily achievable in 384-well plates, roughly ten-fold smaller than reaction volumes typically recommended in commercial MDA kits (Rinke et al. 2014; Tsementzi et al. 2016; Sobol and Kaster 2023).

2.8 Sequencing Library Construction

Whole-genome amplification produces sufficient DNA quantities for constructing sequencing libraries. Sequencing technologies and library construction protocols evolve rapidly, and many options are currently available for single-cell genomic studies. However, long-read sequencing approaches are currently incompatible with de novo genome assembly of single-cell DNA generated by MDA. The hyperbranched nature of MDA results in self-priming reactions that join together noncontiguous genomic regions, and these artifactual "chimeric" junctions occur every 10–20 kb on average (Zhang et al. 2006; Rodrigue et al. 2009; Lasken and Stockwell 2007). Thus, a significant fraction of any long reads in a library will contain errors that appear like genome rearrangements that interfere with contig assembly (unpublished data). In contrast, the impact of short reads spanning a chimeric junction between distant genomic regions is counterbalanced by correct reads derived from non-chimeric copies of these regions (Rodrigue et al. 2009; Chitsaz et al. 2011). Both ligation-based and transposon-based methods are viable options for generating short-read libraries. Finally, dual-unique indexes are critical for short-read sequencing since they mitigate various problems with molecular index hopping and misassignments observed on some short-read sequencing platforms (MacConaill et al. 2018).

2.9 Conclusions

Although laboratory workflows for microbial single-cell genomics can be challenging, developing robust processes for generating SAGs is possible. We hope this general overview, when paired with detailed protocols such as those found in Rinke et al. 2014, and in other chapters in this book, will provide a blueprint for implementing a SAG pipeline. To reiterate, careful attention to decontaminating reagents, labware, and instruments is critical for success, as is careful handling to avoid reintroducing DNA contamination into reactions. Additionally, new developments in WGA chemistries such as PTA, which is entirely compatible with the process described here, may dramatically improve the quality of single-cell microbial genomes (Bowers et al. 2024). Finally, FACS-free methods also hold great potential for single-cell genomics (Leonaviciene et al. 2020; Aoki et al. 2022), and may someday replace the workflows described above. Hopefully, some combination of improved chemistry and sample processing procedures will further enhance the quality and quantity of microbial single-cell genomes in the near future.

References

Falkowski PG, Fenchel T, Delong EF (2008) The microbial engines that drive Earth's biogeochemical cycles. Science 320:1034–1039

Abhilash PC, Dubey RK, Tripathi V, Gupta VK, Singh HB (2016) Plant growth-promoting microorganisms for environmental sustainability. Trends Biotechnol 34:847–850

Tao F et al (2023) Microbial carbon use efficiency promotes global soil carbon storage. Nature 618:981–985

Keasling JD (2010) Manufacturing molecules through metabolic engineering. Science 330:1355–1358

Rappé MS, Giovannoni SJ (2003) The uncultured microbial majority. Ann Rev Microbiol 57:369–394

Sangwan N, Xia F, Gilbert JA (2016) Recovering complete and draft population genomes from metagenome datasets. Microbiome 4:8

Labonté JM, Swan BK, Poulos B, Luo H (2015) Single-cell genomics-based analysis of virus–host interactions in marine surface bacterioplankton. ISME J 9:2386

Roux S et al (2014) Ecology and evolution of viruses infecting uncultivated SUP05 bacteria as revealed by single-cell- and meta-genomics. elife 3:e03125

Jarett JK et al (2020) Insights into the dynamics between viruses and their hosts in a hot spring microbial mat. ISME J 14:2527–2541

Pachiadaki MG et al (2017) Major role of nitrite-oxidizing bacteria in dark ocean carbon fixation. Science 358:1046–1051
Swan BK et al (2011) Potential for chemolithoautotrophy among ubiquitous bacteria lineages in the dark ocean. Science 333:1296–1300
Bowers RM et al (2022) Dissecting the dominant hot spring microbial populations based on community-wide sampling at single-cell genomic resolution. ISME J 16:1337–1347
Rinke C et al (2013) Insights into the phylogeny and coding potential of microbial dark matter. Nature 499:431–437
Pachiadaki MG et al (2019) Charting the complexity of the marine microbiome through single-cell genomics. Cell 179:1623–1635.e11
Engel P, Stepanauskas R, Moran NA (2014) Hidden diversity in honey bee gut symbionts detected by single-cell genomics. PLoS Genet 10:e1004596
Marcy Y et al (2007a) Dissecting biological 'dark matter' with single-cell genetic analysis of rare and uncultivated TM7 microbes from the human mouth. Proc Natl Acad Sci 104:11889–11894
Zheng W et al (2022) High-throughput, single-microbe genomics with strain resolution, applied to a human gut microbiome. Science 376:eabm1483
Rinke C et al (2014) Obtaining genomes from uncultivated environmental microorganisms using FACS–based single-cell genomics. Nat Protoc 9:1038–1048
Woyke T et al (2010) One bacterial cell, one complete genome. PLoS One 5:e10314
Zhang K et al (2006) Sequencing genomes from single cells by polymerase cloning. Nat Biotechnol 24:680–686
Blainey PC, Mosier AC, Potanina A, Francis CA, Quake SR (2011) Genome of a low-salinity ammonia-oxidizing archaeon determined by single-cell and metagenomic analysis. PLoS One 6:e16626
Kaster A-K, Sobol MS (2020) Microbial single-cell omics: the crux of the matter. Appl Microbiol Biotechnol 104:8209–8220
Rodrigue S et al (2009) Whole genome amplification and de novo assembly of single bacterial cells. PLoS One 4:e6864
He J, Du S, Tan X, Arefin A, Han CS (2016) Improved lysis of single bacterial cells by a modified alkaline-thermal shock procedure. BioTechniques 60:129–135
Nishikawa Y et al (2022) Validation of the application of gel beads-based single-cell genome sequencing platform to soil and seawater. ISME Commun 2:92
Leonaviciene G, Leonavicius K, Meskys R, Mazutis L (2020) Multi-step processing of single cells using semi-permeable capsules. Lab Chip 20:4052–4062
Chijiiwa R et al (2020) Single-cell genomics of uncultured bacteria reveals dietary fiber responders in the mouse gut microbiota. Microbiome 8:5
Aoki H et al. (2021) Agarose gel microcapsules enable easy-to-prepare, picolitre-scale single-cell genomics, yielding near-complete genome sequences. https://doi.org/10.21203/RS.3.RS-147972/V1
Shapiro H (2003a) Chapter 7.2: Intrinsic cellular parameters. In: Shapiro H (ed) Practical flow cytometry, 4th edn. Wiley, pp 285–293
Stepanauskas R et al (2017) Improved genome recovery and integrated cell-size analyses of individual uncultured microbial cells and viral particles. Nat Commun 8:84
Kashtan N et al (2014) Single-cell genomics reveals hundreds of coexisting subpopulations in wild Prochlorococcus. Science 344:416–420
Wu Y-T et al (2021) Single-cell genomics-based analysis reveals a vital ecological role of Thiocapsa sp. LSW in the meromictic Lake Shunet, Siberia. Microb Genom 7
Nakayama T et al (2019) Single-cell genomics unveiled a cryptic cyanobacterial lineage with a worldwide distribution hidden by a dinoflagellate host. Proc Natl Acad Sci USA 116:15973–15978
Birzu G et al (2023) Hybridization breaks species barriers in long-term coevolution of a cyanobacterial population. eLife. https://doi.org/10.7554/eLife.90849.1
Grieb A et al (2020) A pipeline for targeted metagenomics of environmental bacteria. Microbiome 8:21

Yilmaz S, Haroon MF, Rabkin BA, Tyson GW, Hugenholtz P (2010) Fixation-free fluorescence in situ hybridization for targeted enrichment of microbial populations. ISME J 4:1352–1356
Podar M et al (2007) Targeted access to the genomes of low-abundance organisms in complex microbial communities. Appl Environ Microbiol 73:3205–3214
Munson-McGee JH et al (2022) Decoupling of respiration rates and abundance in marine prokaryoplankton. Nature 612:764–770
Couradeau E et al (2019) Probing the active fraction of soil microbiomes using BONCAT-FACS. Nat Commun 10:2770
Hatzenpichler R, Connon SA (2016) Visualizing in situ translational activity for identifying and sorting slow-growing archaeal−bacterial consortia. Proc Natl Acad Sci U S A 113(28):E4069–E4078
Needham DM et al (2019) A distinct lineage of giant viruses brings a rhodopsin photosystem to unicellular marine predators. Proc Natl Acad Sci USA 116:20574–20583
Lasken RS (2007) Single-cell genomic sequencing using multiple displacement amplification. Curr Opin Microbiol 10:510–516
Dean FB, Nelson JR, Giesler TL, Lasken RS (2001) Rapid amplification of plasmid and phage DNA using Phi29 DNA polymerase and multiply-primed rolling circle amplification. Genome Res 11:1095–1099
Raghunathan A et al (2005) Genomic DNA amplification from a single bacterium. Appl Environ Microbiol 71:3342–3347
Marcy Y et al (2007b) Nanoliter reactors improve multiple displacement amplification of genomes from single cells. PLoS Genet 3:1702–1708
Clingenpeel S, Clum A, Schwientek P, Rinke C, Woyke T (2014a) Reconstructing each cell's genome within complex microbial communities–dream or reality? Front Microbiol 5:771
Chen C et al (2017) Single-cell whole-genome analyses by Linear Amplification via Transposon Insertion (LIANTI). Science 356:189–194
de Bourcy CFA et al (2014) A quantitative comparison of single-cell whole genome amplification methods. PLoS One 9:e105585
Lan F et al (2024) Massively parallel single-cell sequencing of diverse microbial populations. Nat Methods 21:228–235
Wang G et al (2023) High-throughput generic single-entity sequencing using droplet microfluidics. bioRxiv 2023.08.15.549386. https://doi.org/10.1101/2023.08.15.549386
Abate A et al (2023) Microbiome single cell atlases generated with a commercial instrument. Res Sq. https://doi.org/10.21203/rs.3.rs-3253785/v1
Bowers RM et al (2024) scMicrobe PTA: near complete genomes from single bacterial cells. ISME Communications 4(1):ycae085. https://doi.org/10.1093/ismeco/ycae085
Gonzalez-Pena V et al (2021) Accurate genomic variant detection in single cells with primary template-directed amplification. Proc Natl Acad Sci 118:e2024176118
Clingenpeel S, Schwientek P, Hugenholtz P, Woyke T (2014b) Effects of sample treatments on genome recovery via single-cell genomics. ISME J 8:2546–2549
Marie D, Rigaut-Jalabert F, Vaulot D (2014) An improved protocol for flow cytometry analysis of phytoplankton cultures and natural samples. Cytometry A 85:962–968
Swan BK et al (2013) Prevalent genome streamlining and latitudinal divergence of planktonic bacteria in the surface ocean. Proc Natl Acad Sci USA 110:11463–11468
Gasol JM, del Giorgio PA (2000) Using flow cytometry for counting natural planktonic bacteria and understanding the structure of planktonic bacterial communities. Sci Mar 64:197–224
Woyke T et al (2011) Decontamination of MDA reagents for single cell whole genome amplification. PLoS One 6:e26161
Shapiro H (2003b) Chapter 4.6: Flow systems. In: Shapiro H (ed) Practical flow cytometry, 4th edn. Wiley, pp 166–180
Schweitzer HD et al (2022) Subsurface hydrocarbon degradation strategies in low- and high-sulfate coal seam communities identified with activity-based metagenomics. NPJ Biofilms Microbiomes 8:7

Shapiro H (2003c) Chapter 7.1: Physical parameters and their uses. In: Shapiro H (ed) Practical flow cytometry, 4th edn. Wiley, pp 273–284

Azam F, Hodson RE (1977) Size distribution and activity of marine microheterotrophs1. Limnol Oceanogr 22:492–501

National Research Council (US) Steering Group for the Workshop on Size Limits of Very Small Microorganisms. Diminutive cells in the oceans—unanswered questions. (National Academies Press (US), 1999)

Levin PA, Angert ER (2015) Small but mighty: cell size and bacteria. Cold Spring Harb Perspect Biol 7:a019216

Morono Y, Terada T, Hoshino T, Inagaki F (2014) Hot-alkaline DNA extraction method for deep-subseafloor archaeal communities. Appl Environ Microbiol 80:1985–1994

Kvist T, Ahring BK, Lasken RS, Westermann P (2007) Specific single-cell isolation and genomic amplification of uncultured microorganisms. Appl Microbiol Biotechnol 74:926–935

Stepanauskas R (2012) Single cell genomics: an individual look at microbes. Curr Opin Microbiol 15:613–620

Tsementzi D et al (2016) SAR11 bacteria linked to ocean anoxia and nitrogen loss. Nature 536:179–183

Ciobanu D et al (2022) Protocol for single-cell isolation and genome amplification of environmental microbial eukaryotes for genomic analysis. STAR Protoc 3:100968

Ishoey T, Woyke T, Stepanauskas R, Novotny M, Lasken RS (2008) Genomic sequencing of single microbial cells from environmental samples. Curr Opin Microbiol 11:198–204

Blainey PC, Quake SR (2011) Digital MDA for enumeration of total nucleic acid contamination. Nucleic Acids Res 39:e19

Vollmers J, Wiegand S, Lenk F, Kaster A-K (2022) How clear is our current view on microbial dark matter? (Re-)assessing public MAG & SAG datasets with MDMcleaner. Nucleic Acids Res 50:e76–e76

Sobol MS, Kaster A-K (2023) Back to basics: a simplified improvement to multiple displacement amplification for microbial single-cell genomics. Int J Mol Sci 24

Lasken RS, Stockwell TB (2007) Mechanism of chimera formation during the Multiple Displacement Amplification reaction. BMC Biotechnol 7:19

Chitsaz H et al (2011) Efficient de novo assembly of single-cell bacterial genomes from short-read data sets. Nat Biotechnol 29:915–921

MacConaill LE et al (2018) Unique, dual-indexed sequencing adapters with UMIs effectively eliminate index cross-talk and significantly improve sensitivity of massively parallel sequencing. BMC Genomics 19:30

Aoki H et al (2022) Agarose gel microcapsules enable easy-to-prepare, picolitre-scale, single-cell genomics, yielding high-coverage genome sequences. Sci Rep 12:17014

Characterization of Microbial Diversity and Mobile Genetics Elements Through Single-Cell Genomics

3

Alicia Ortiz-Maiques and Mária Džunková

Abstract

The ongoing advancement of cultivation-independent techniques has facilitated the discovery and characterization of various microbial species and their mobile genetic elements (MGEs). Single-cell approaches, which address the limitations of metagenomics, offer invaluable insights into microbial diversity, various genetic elements, and their interactions within the microbiome. In this chapter, we introduce innovative methodologies in single-cell genomics, including a microfluidics technology and various whole-genome amplification techniques that have transformed the resolution of microbial genome sequencing. We present two high-throughput workflows for single-cell genomic analysis: the label—sort—encapsulate—amplify (LSEA) strategy and the encapsulate—amplify—label—sort (EALS) strategy. These approaches are designed for the targeted isolation and sequencing of individual single cells, enabling a comprehensive study of bacterial genomes and their MGEs, thereby enhancing our understanding of microbial ecology.

Keywords

Microbial single-cell genomics · Microfluidics · Atrandi · Mobile genetic elements · Capsule-PCR

A. Ortiz-Maiques · M. Džunková (✉)
Institute for Integrative Systems Biology (I2SysBio), University of Valencia and Spanish National Research Council (CSIC), Valencia, Spain
e-mail: alicia.ortiz@uv.es; maria.dzunkova@uv.es

C. Rinke, M. Džunková (eds.), *Single-Cell Omics in Microbiome Research*, The Microbiomes of Humans, Animals, Plants, and the Environment 7,
https://doi.org/10.1007/978-3-032-07527-7_3

3.1 Introduction

Traditional microbiological techniques focus on isolating and cultivating microorganisms to describe their traits, including metabolism, growth, or various physiological states specific to certain organisms (Lewis et al. 2021). However, recent advancements in metagenomic sequencing have shown that the fraction of microbial taxa in natural environments that can be cultured under standard laboratory conditions is remarkably low, likely well below 1% (Steen et al. 2019), suggesting an even greater gap between the vast microbial diversity and the limited number that can be cultured. Difficulties in microorganism cultivation often stem from the challenge of recreating optimal growth conditions, which may require extreme environments or unknown nutrients (Lewis et al. 2021).

Despite advancements in classical microbiology, the development of various technologies has been crucial in characterizing what is known as "microbial dark matter". This concept includes previously unidentified bacterial and archaeal taxa, as well as viruses representing the most abundant molecular entities across all environments and a major reservoir of genetic diversity (Paez-Espino et al. 2016). Consequently, the implementation of cultivation-independent methods, such as metagenomics and single-cell genomics (Fig. 3.1), has dramatically transformed the field of microbiology, facilitated by progress in single-cell separation technologies,

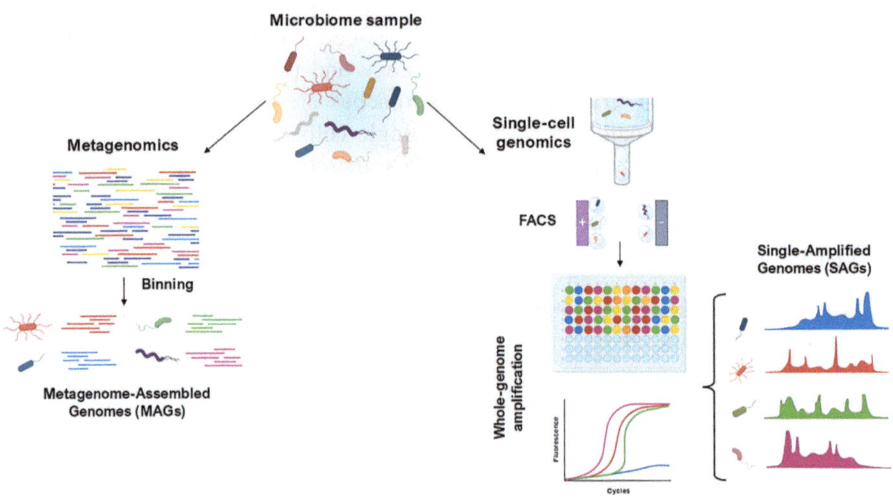

Fig. 3.1 Workflow representation of genomic techniques for microbiome studies. Metagenomics and single-cell genomics are two key approaches for studying microbial genomes. Metagenomics involves sequencing the entire genetic material of an environmental sample, followed by read classification and binning to reconstruct metagenome-assembled genomes (MAGs). In contrast, single-cell genomics employs techniques like FACS to isolate individual cells, enabling genome sequencing at the single-cell level and generating single-amplified genomes (SAGs). (Created with BioRender)

whole genome amplification chemistry, sequencing technologies, and data analysis approaches.

This chapter introduces single-cell genomics methodologies and their advantages over classical metagenomics in understanding microbial ecology and evolution. We describe emerging high-throughput single-cell approaches in this field, including microfluidic technology and cutting-edge genome amplification methods. Finally, we will discuss their applications and strategies in the study of complex microbial communities.

3.2 Single-Cell Genomics Methods for Microbiome Studies

3.2.1 From Metagenomics to a Single-Cell Approach

Metagenomics is now firmly established as a distinct scientific discipline for analyzing complex environmental microbial samples (Berg et al. 2020). It examines all the microbial genetic material found in an environment and allows characterization of the taxonomic diversity of microbial communities, assesses their functional potential, and identifies genome sequences of various bacterial species as well as molecular entities, such as previously unknown viruses or phages (Quince et al. 2017). Consequently, metagenomics has significantly reshaped the concept and structure of the Tree of Life (Lewis et al. 2021).

Metagenomics enables the reconstruction of metagenome-assembled genomes (MAGs), which include prokaryotic and viral genomes within complex samples. However, these techniques come with certain limitations (Fig. 3.2). Metagenomics studies often generate an overwhelming amount of data, making bioinformatics tools essential for processing results efficiently, starting with reconstructing genomic sequences through read assembly, followed by binning of the resulting contigs into MAGs (Alneberg et al. 2018; Garfias-Gallegos et al. 2022).

The binning strategy is crucial in identifying various intraspecific genetic variations of microorganisms, which may be overlooked during data processing. Typically, the MAGs recovered from microbial communities represent a consensus genome from a closely related population rather than individual genomes (Alneberg et al. 2018). Further issues include contamination and low completeness of the resulting MAGs (Bowers et al. 2017). Bin contamination occurs when sequences from different organisms are mistakenly grouped, due to assembly or binning algorithms, leading to chimeric genomes and misleading biological interpretations. Low genome completeness results from insufficient contigs placed in a bin, often due to assembly challenges arising from the high complexity of microbial communities. Although new bioinformatic tools are being developed to improve genome quality (Yang et al. 2021), this challenge continues to hinder the accurate reconstruction of individual genomes and, consequently, the description of new bacterial species based on their genome sequences (Hedlund et al. 2022). Furthermore, binning methods also often struggle to associate mobile genetic elements (MGEs) with the genomes into which they were integrated (Maguire et al. 2020). Some bioinformatics tools that employ

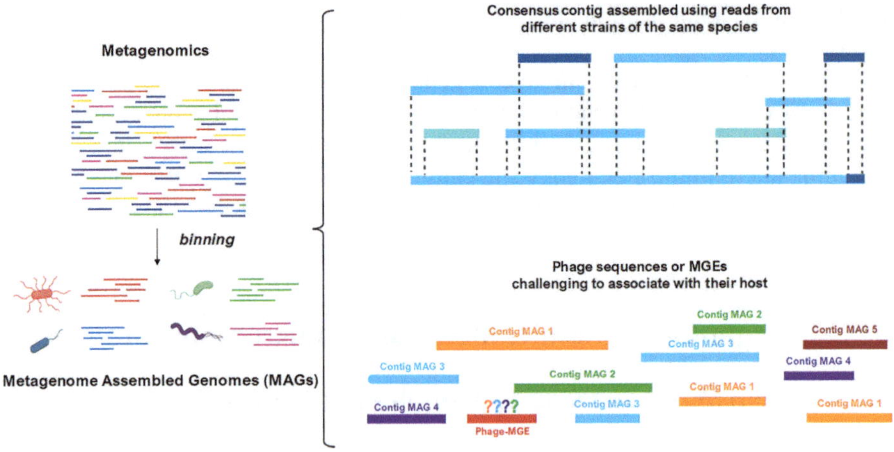

Fig. 3.2 Limitations of metagenomics in microbiome studies. Genomes are reconstructed as consensus sequences representing a given microbial population during metagenomic data processing. However, this approach can obscure valuable information about the genetic variability within species. Another significant challenge is the limited understanding of the genetic signatures that connect mobile genetic elements (MGEs) or phages with their hosts. (Created with BioRender)

network-based and co-occurrence modeling have enhanced the prediction of interactions between phages and their hosts, addressing previous challenges in identifying MGEs (Edwards et al. 2016), but it is still challenging.

MGEs, which include plasmids, genomic islands, or fragments from phage genomes, encode enzymes or proteins responsible for DNA movement within a genome or between different bacteria (Frost et al. 2005). Introducing these elements into the bacterial genome through horizontal gene transfer drives diversification and evolution in microbial populations (Weltzer and Wall 2023). Therefore, studying MGEs is essential for characterizing the genetic diversity of microorganisms and understanding phage–host interactions in natural environments. However, investigating these relationships remains challenging, as the genetic signatures associated with previous infections are scarce across the tree of life, which impedes predictions of potential phage–host pairs using bioinformatics tools in many bacterial taxa (Paez-Espino et al. 2016).

In contrast to metagenomics, single-cell genome sequencing provides a detailed characterization of complex microbial communities at the single-cell resolution. Single-amplified genome (SAG) recovery begins with isolating individual cells from the environmental sample (Woyke et al. 2017) (Fig. 3.1). Once isolated, each cell undergoes DNA extraction, whole-genome amplification, sequencing, and assembly of the genome of each individual cell (Grün and van Oudenaarden 2015).

Single-cell analysis helps overcome several inherent struggles of metagenomics, including difficulties characterizing intraspecific genetic diversity and establishing relationships between MGEs or phage sequences and their hosts (Alneberg et al. 2018). By working at the single-cell level, researchers can ensure that the resulting reads originate from the genome of an individual cell, along with any MGEs it may carry.

However, implementing broader microbial single-cell strategies can be challenging. The small reagent volumes used for individual cells sorted by fluorescence-activated cell sorter (FACS) into 96- or 384-well plates and the high costs associated with creating a separate sequencing library for each cell (often resulting in low-quality data) pose significant limitations. These challenges have driven the development of new strategies and methodologies aimed at enhancing data quality while reducing costs and the labor-intensive processes of the technique.

3.2.2 Challenges of Single-Cell Genomics

3.2.2.1 Chemistry Limitations

The growing interest in analyzing genetic material at single-cell resolution has driven significant advancements in sequencing technology. As a result, the sensitivity of sequencing processes has markedly increased, allowing for successful analysis with smaller quantities of DNA per experimental run (Wang and Navin 2015). However, the genetic material content of an individual cell (either eukaryotic or prokaryotic) does not match this need, so a critical step in single-cell experiments is the whole-genome amplification (WGA) procedure to achieve the required quantity of DNA (Gawad et al. 2016). Recent technological improvements, such as microfluidic-based WGA approaches, have further optimized this step by distributing and amplifying single-cell genome fragments in hundreds of thousands of droplets, obtaining high-quality SAGs by reducing amplification bias, and increasing genome recovery rates compared to conventional in-tube WGA (Hosokawa et al. 2017).

One of the most commonly used methods for whole-genome amplification is Multiple Displacement Amplification (MDA) (Fig. 3.3) (Ishoey et al. 2008). MDA uses the highly processive Φ29 DNA polymerase, which can synthesize DNA while displacing the existing strands and exhibits a relatively low error rate (Ahsanuddin et al. 2017). This polymerase amplifies DNA fragments at an optimal temperature of approximately 30 °C, using short random primers that hybridize to different genomic sequences (Grün and van Oudenaarden 2015).

Despite its advantages, MDA presents several limitations due to its exponential amplification nature. The reaction can lead to branched DNA structures, which introduce significant biases in genome representation (Lu et al. 2023). Consequently, genomic regions preferentially amplified by Φ29 will be detected in extremely high proportions, whereas others may be underrepresented or absent. If MDA is applied to small bulks of cells (mini-metagenomes), this amplification bias will distort the original abundance of species in a complex sample, leading to the loss of sequence information from low-abundance organisms (Quince et al. 2017). Additionally, branched amplification can result in artifacts, such as chimeric DNA fragments derived from non-adjacent genomic regions, complicating genome assembly (Lu et al. 2023). To address these limitations, some studies have focused on using thermostable mutants of the Φ29 polymerase to improve genome recovery from individual microbial cells while maintaining comparable fidelity. This approach has

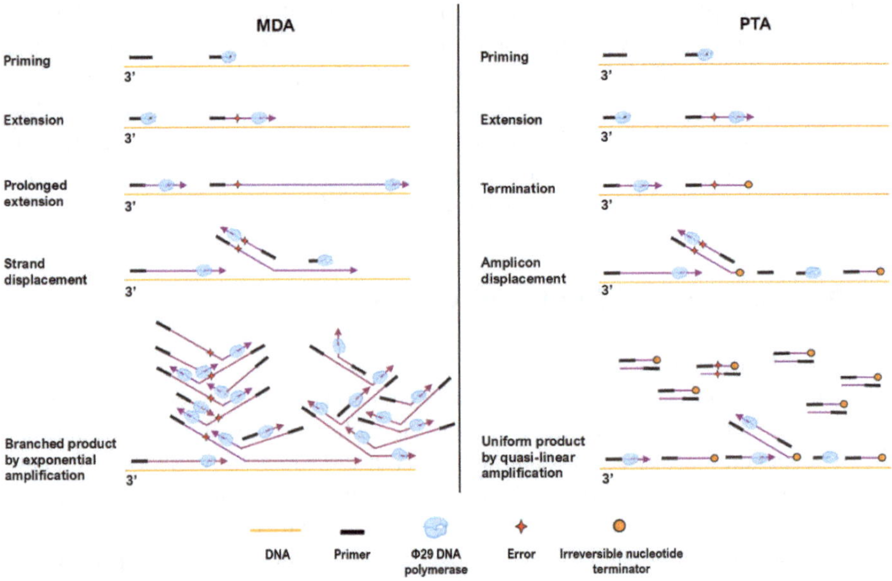

Fig. 3.3 Comparison of whole genome amplification methods. Multiple displacement amplification (MDA) and primary template-directed amplification (PTA). Each method utilizes distinct priming, DNA synthesis, and amplification strategies, resulting in PTA providing improved genome coverage and reduced amplification bias. The legend highlights essential components, including DNA templates, primers, polymerases, and termination bases. (Created with BioRender)

shown enhanced efficacy in recovering G + C-rich sequences, often more common in prokaryotes (Stepanauskas et al. 2017).

Alternative strategies to MDA have been designed, including the primary template-directed amplification (PTA) (Fig. 3.3) (Gonzalez-Pena et al. 2021). This method retains the advantageous properties of Φ29 DNA polymerase while incorporating modified nucleotides resistant to the enzyme's 5′ exonuclease activity. As a result, PTA generates smaller double-stranded DNA amplicons, promoting a nearly linear amplification process. This approach significantly reduces error propagation and ensures more uniform and comprehensive genome coverage.

The PTA method has already been successfully applied in sequencing human neural cells, demonstrating that somatic point mutations and small insertion/deletion (indels) mutations in regulatory expression regions accumulate over time, impacting neuronal integrity (Luquette et al. 2022). However, its application in prokaryotic cells remains relatively unexplored, with only one study investigating its potential in bacterial sequencing (Bowers et al. 2024). In summary, PTA is a promising but still emerging technique for microbiome research. Expanding its application in bacterial and viral genomics could offer new insights into microbial evolution and functional diversity, which may lead to innovative uses in environmental and clinical microbiology.

3.2.2.2 Throughput

The single-cell methods require the physical isolation of individual cells. One of the most commonly used techniques for this purpose is FACS, which operates by flow cytometry principles (Woyke et al. 2017). This approach usually involves prior fluorescence labeling of the target cells, such as fluorescence in situ hybridization (FISH), which employs fluorescent probes complementary to specific sequences in the genome of interest (Pereira et al. 2022). FISH has gained immense popularity due to its ability to provide high specificity, sensitivity, and multiplexing potential, allowing for the simultaneous detection of multiple genetic targets in a single cell. The FISH-FACS protocol allows the identification of target cells based on their fluorescence emission and other cellular characteristics, such as size or granularity. Once identified, the selected cells are individually sorted, typically into a multi-well plate where the subsequent procedures can be performed (Lawrence et al. 2022).

Efforts to enhance throughput in single-cell studies have led to the development of new isolation technologies, such as microfluidics, which have become increasingly popular (Grün and van Oudenaarden 2015). Traditional methods, such as FACS-based workflows, face significant challenges related to throughput, mainly due to difficulties in distributing reagents across multiple wells. Achieving precise and efficient liquid handling often requires expensive robotic systems, posing a significant cost barrier for many laboratories. In contrast, microfluidics offers a miniaturized and cost-effective solution, allowing researchers to separate and isolate thousands of cells into droplets or compartments of varying types. These can be processed and analyzed later in a controlled environment (Lan et al. 2023). As a result, microfluidics facilitates the simultaneous study of thousands of cells within a single experiment.

Among the principal challenges of implementing microfluidic platforms in microbial single-cell genomics is the need to facilitate the exchange of reagents and temperature changes during the diverse steps of single-cell DNA extraction and WGA (Wang et al. 2023). Capsules formed by microfluidic devices are often made from unstable polymers that can either be impermeable to essential reagents or unable to withstand specific conditions, such as the high temperatures required for some molecular reactions (Duran et al. 2022).

These limitations have prompted innovation, including semi-permeable hydrogel development for single-cell encapsulation, which has been successfully applied in various biochemical assays (Leonaviciene et al. 2020). Among the most innovative hydrogel capsule technologies available, it is distributed by Atrandi Bioscience (https://atrandi.com/), a company specializing in microfluidic technology. Their system encapsulates single cells within a semi-permeable hydrogel composed of dextrans and modified polyethylene glycol acrylate, forming a biphasic aqueous system (Leonaviciene et al. 2020). Incorporating hydrogels in encapsulation protocols has been a game-changer, as it maintains cell viability and supports cellular function throughout the experimental processes. These semi-permeable capsules (SPCs) made from hydrogel allow precise control of the microenvironments surrounding isolated cells. This control is crucial for conducting various molecular reactions on the cell's genetic material, including DNA extraction and amplification procedures, such as PCR and whole-genome amplification. These reactions are performed with thousands of encapsulated single cells within a single tube, streamlining laboratory workflows. This means that the struggles associated with the distribution of the small reaction volumes across single-cells sorted into 96 or 384 well plates are eliminated, improving the efficiency and throughput of experimental design.

Additionally, Atrandi has developed a unique barcoding system for the genetic material from single cells to facilitate multiplexing. This split-and-pool barcoding system functions similarly to the MicroSPLiT protocol (Kuchina et al. 2021), enabling the simultaneous processing and sequencing of tens of thousands of genomes in a single run.

3.3 High-Throughput Single-Cell Genomics Workflow

This section describes two experimental approaches using microfluidics to conduct high-throughput single-cell analysis of environmental samples. The two approaches differ in how fluorescent labeling specific to the target cell population is applied to the environmental sample (Fig. 3.4).

Both the label—sort—encapsulate—amplify (LSEA) and the encapsulate—amplify—label—sort (EALS) approaches share initial steps. These include fixing the environmental microbial sample and the fluorescent labelling of cells to collect thousands of cells by FACS sorting. These cells are subsequently encapsulated by the microfluidic device and processed as single cells in a single tube. Encapsulated cells are lysed and then their genome is amplified by a whole-amplification method. In the LSEA method, the workflow continues with barcoding the amplified fragments from each capsule using a split-and-pool strategy. This enables multiplexed sequencing and downstream read assignments through bioinformatics tools like CleanBar. In contrast, the EALS approach involves a capsule-PCR process that uses fluorescently labeled primers to detect specific genetic elements. The fluorescent capsules are then identified and sorted using a COPAS cytometer, allowing for the targeted isolation of genomes of interest.

The choice of the workflow should be based on the labeling approach. Labeling by activity-based probes, such as BONCAT (Reichart et al. 2020) or fluorescent pantetheine probes (Džunková et al. 2023), is fully compatible with the LSEA. Furthermore, the classical FISH labeling, which targets specific bacterial lineages via fluorescent 16S rRNA probes, is also suitable for the LSEA approach. Nevertheless, special attention must focus on elimination of aggregated cells before encapsulation. FISH reagents tend to aggregate cells in liquid suspension; however, doublet discrimination and size-based gating in FACS analysis can ensure that only true bacterial single cells are sorted and forwarded to the microfluidics encapsulation step. Still, FISH reagents may inadvertently cause small particles, such as viruses, to attach to the cell surface, potentially remaining undetected even when FACS size restrictions and doublet discrimination are applied. As a result, the LSEA

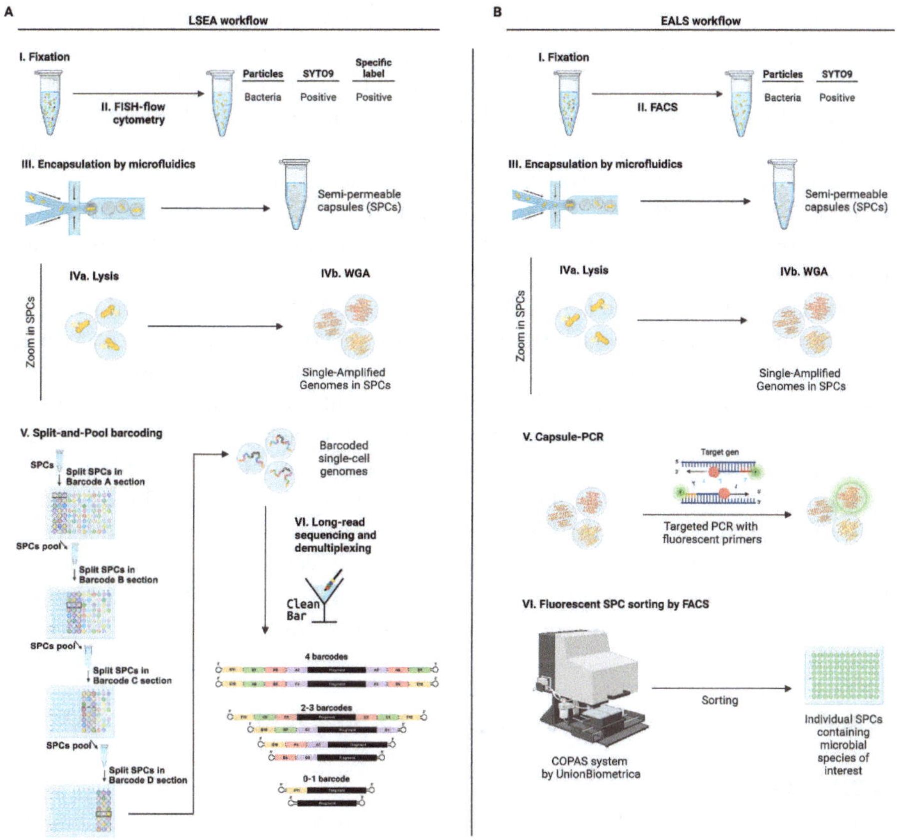

Fig. 3.4 Overview of single-cell genomics approaches. Shown are the steps involved in two workflows, (**a**) label—sort—encapsulate—amplify (LSEA) and (**b**) Encapsulate—amplify—label—sort (EALS). Both start with similar steps consisting of the fixation of an environmental microbial sample, separation and collection of the cells by flow cytometry and cell encapsulation, followed by cell lysis and whole genome amplification. The LSEA approach (**a**) proceeds with the distinct labeling of the amplified fragments from each capsule, utilizing a split-and-pool barcoding procedure. To analyze the sequencing results of this multiplexed experiment, bioinformatics tools like CleanBar are employed to classify the reads based on the barcode associated with each fragment. On the other hand, the EALS approach (**b**) follows with the labeling of capsules containing a specific genetic element by a capsule-PCR using modified fluorescent primers. These labeled capsules can be detected by the COPAS cytometer and separated individually in a multi-well plate for subsequent processing. (Created with BioRender)

approach is less suitable for studying viral infections in specific bacterial lineages targeted by these probes.

In contrast, the EALS approach enables the study of natural phage infections in environmental samples without the complications caused by particle aggregation due to the application of fluorescent probes. In this approach, natural bacterial populations are encapsulated without chemical treatment, ensuring that attached particles are captured in their natural state rather than as artifacts of aggregation caused by chemical treatment involved in the fluorescent labelling process. After encapsulation, WGA is performed, followed by applying specific fluorescent primers to label and sort the target bacterial lineage that may harbor a new phage species or a yet-undiscovered MGE. Conversely, fluorescent primers can also be used to target a specific virus or MGE, enabling the identification of its previously unknown bacterial hosts. This method is particularly advantageous in laboratories where the same primers, but without fluorescent tags, are routinely used for detecting the microbe of interest (either bacteria or viruses) in environmental samples through traditional PCR.

The choice between the LSEA (Fig. 3.4a) and EALS (Fig. 3.4b) workflows can also depend on the desired sequencing strategy. For projects aiming to sequence a large number of cells in a single run, LSEA is the most suitable approach, as it enables multiplexing across 10,000 cells. Sufficient sequencing capacity must be allocated to achieve adequate genome coverage for each cell. In contrast, EALS may be preferable for researchers looking to sequence only a few cells from the target population or when the target population is too small to yield thousands of cells during FACS sorting, the first step of the LSEA method. In EALS, fluorescent capsules are collected into separate wells and processed as individual sequencing libraries, eliminating the need for the split-and-pool barcoding kit used in the high-throughput LSEA approach. This also makes EALS a more cost-effective option.

3.3.1 Label—Sort—Encapsulate—Amplify (LSEA)

3.3.1.1 Cell Fixation

Before cell fixation and labeling, it is important to properly prepare the sample to ensure optimal cell concentration and minimize background noise. The sample preparation process varies depending on the type of sample being analyzed. For environmental samples, it is essential to clean and concentrate them to isolate the bacterial fraction. For instance, when dealing with fecal samples, a small amount can be resuspended in phosphate-buffered saline (PBS) and allowed to settle. This settling process helps remove larger particles that could interfere with further analysis. After settling, the supernatant containing the bacterial fraction is recovered and centrifuged to wash and concentrate the microbial sample before proceeding with labeling.

Microbial sample fixation is critical for neutralizing potential biosafety risks associated with unknown microbial content while preserving cellular structure. Initially, the environmental sample is filtered using a 5.0-micrometer membrane to remove larger particles and aggregates, ensuring that only the microbial fraction will be processed. Methacarn, a fixative composed of methanol, concentrated acetic acid, and chloroform in a 6:1:3 ratio, is used due to its superior preservation of genomic DNA and other cellular components rather than formaldehyde (Shibutani and Uneyama 2005; Shibutani et al. 2000). Unlike formaldehyde, methacarn ensures higher-quality nucleic acid recovery for subsequent genomic analyses; it also prevents excessive protein cross-linking, allowing for efficient genetic material extraction and maintaining better cellular and nuclear morphology, essential for single-cell studies. Furthermore, methacarn reduces contamination risks and experimental artifacts, making it a more reliable fixative for microbiome and single-cell genomic research (Clingenpeel et al. 2014).

The filtered sample is concentrated by centrifugation at 10,000 x g for 5 minutes. The resulting pellet is incubated in 1 mL of methacarn for 30 minutes. To remove residual fixatives that may interfere with downstream reactions, several washes with phosphate-buffered saline (PBS) are performed before proceeding to subsequent steps.

3.3.1.2 Fluorescent Labeling and Sorting

In the next step, FACS is used to identify the microbial populations of interest. The cell size and nucleic acid content corresponding to different fluorescence levels can be determined using nonspecific nucleic acid stains. It is recommended to use these generic nucleic acid stains (such as SYTO9 or SYTO 62) also in combination with specific fluorescent probes (e.g., specific FISH probes) since double labeling helps to identify false positives. For labeling, 1 μL of each dye is added to 1 mL of the cell suspension, followed by incubation in the dark for 30 minutes to preserve fluorescence integrity. Flow cytometry is used to analyze and sort the particles based on fluorescent emission and size. Each collected sample should contain exactly 62,500 particles, as recommended by Atrandi, to obtain the highest yields and the lowest number of multiplets.

3.3.1.3 Microfluidic Encapsulation

Microfluidic encapsulation is applied to the cells sorted by FACS. The Atrandi FLUX microfluidic device and the SPCs Generation Kit (CKP-BARK1, version 6) generate semi-permeable capsules around individual cells. The working solutions are prepared as follows:

1. *Working Core Solution (WCS):* 50 μL of 2x Core Reagent, 1 μL of dithiothreitol (DTT), 12.5 μL of photoinitiator, and 36.5 μL of the 62,500 microbial particles sample.
2. *Working Shell Solution (WSS):* 50 μL of 2x Shell Solution with 50 μL of PBS.

These solutions are loaded into the microfluidic chip well (300 μL of capsule stabilization oil, 100 μL WSS, and 100 μL WCS). Encapsulation and cross-linking occur within the FLUX system for 30 minutes. The emulsion is transferred to a 1.5 mL tube, where 500 μL of capsule wash buffer and 500 μL of emulsion breaker are added, mixed, and centrifuged. The capsules (SPCs) are then isolated at the interface between the aqueous phase and the emulsion breaker. The SPCs are then resuspended with the aqueous layer and washed four times with capsule wash buffer to remove residual oils. SPCs are then fixed by adding 900 μL of cold methanol to 100 μL of SPCs while agitating. Fixed samples can be stored at −20 °C for long-term before lysis and WGA.

3.3.1.4 Cell Lysis and Whole-Genome Amplification

SPCs are washed 5 times with 1x wash buffer and resuspended in a final volume of 500 μL. Then, 500 μL of lysis buffer (2x alkaline lysis buffer and 100 μL of 1 M DTT) is added and mixed by vortexing. After incubation for 15 minutes at room temperature, five washes with neutralization buffer are performed, followed by five washes with 1x Wash Buffer. WGA is performed using a reaction mix containing 150 μL of SPCs, 97.5 μL of nuclease-free water, 37.5 μL of 10x WGA reaction buffer, 37.5 μL of dNTP mix, 18.75 μL of primer mix, 3.75 μL of 0.1 M de DTT, 3.75 μL of 10% Pluronic F-68, 7.5 μL of WGA enhancer, and 18.75 μL of WGA polymerase. The total reaction mix volume is distributed into 5 PCR tubes with 75 μL each. The reaction is incubated at 45 °C for 1 hour, followed by enzyme inactivation at 65 °C for 10 minutes. Amplification efficiency is assessed via fluorescence microscopy (e.g., Leica Thunder DMI8) using SYTO9 staining. For this, 1 μL of a 1:10 dilution of the nucleic acid fluorophore SYTO9 is added to a 1:10 dilution of treated SPCs, in a final volume of 50 μL. Capsules that have successfully amplified their genome content will appear fluorescent.

3.3.1.5 Split-Pool Method for Barcoding

After the WGA, the debranching and end preparation reactions are performed according to the single-microbe DNA barcoding protocol (version 3). The debranching reaction mix consists of 150 μL of SPCs, 30 μL of 10x debranching buffer, 15 μL of debranching enzyme, and 105 μL of nuclease-free water (NFW), incubated in a thermomixer at 37 °C for 1 hour, followed by three washes with 1x Wash Buffer. The end preparation mix involves 21 μL of End Prep Reaction Buffer, 9 μL of End Prep Enzyme, and 150 μL of SPCs, incubated at 25 °C for 30 minutes and then 65 °C for 30 minutes.

Following further washing, the split-and-pool barcoding strategy uses Atrandi's 96-well plate system. A master mix consisting of 150 μL SPCs, 60 μL ligation buffer, 20 μL ligation enzyme, and 70 μL NFW is distributed across 24 wells of the A section and incubated at room temperature for 15 minutes. Following this, 40 μL Stop Buffer is added to each well used and incubated for an additional 5 minutes. After collecting and washing the SPCs treated in the 24 wells, the pooled capsules are again distributed in the neighboring 24 wells of the same plate. These steps are repeated in the four sections defined in the 96-well plate system of Atrandi to achieve final reads with a barcode consisting of four 8-bp sequences, unique for each capsule content.

Finally, DNA is obtained by decomposing capsules using 4 μL of Release Reagent per 100 μL of SPCs. After a 5-minute incubation, DNA is purified using the AMPure XP protocol using the 0.5x beads to sample ratio to remove non-ligated barcodes and short fragments.

3.3.1.6 Genomic Sequencing and Bioinformatic Analysis

For high-throughput sequencing, long-read sequencing technology is required to ensure accurate and comprehensive genome reconstruction. Library preparation can be performed using the PacBio HiFi protocol, which enables high-fidelity sequencing of long reads. DNA concentration is assessed using a Qubit fluorometer, and sequencing library preparation follows the manufacturer's instructions, specific to the chosen technology. It is important to note that if Illumina sequencing is chosen instead, be aware that adjustments to the library preparation protocol will be necessary to accommodate this different technology. These adjustments involve utilizing specific adapters and an additional amplification step to ensure enough DNA is available for sequencing, tailored to the requirements of the Illumina platform.

Given that the samples contain barcoded reads, a demultiplexing procedure is required. Our laboratory has developed a bioinformatics tool called CleanBar (Arnau et al. 2025), designed specifically for decoding single-cell sequencing reads (https://github.com/tbcgit/cleanbar). Initially, raw reads are filtered based on quality scores, and barcode sequences are identified and assigned to classify each read, ensuring correct assignment to individual cells. Demultiplexing will yield thousands of separate FASTQ files that can be subsequently used for single-genome assemblies, taxonomic classification, functional annotation, detection of viruses and MGEs, and comparative genomic studies, facilitating the exploration of complex environmental microbiomes at a single-cell level.

3.3.2 Encapsulate—Amplify—Label—Sort (EALS)

The EALS approach is suitable for research projects that target specific genes in a microbial community using PCR. The gene of interest may be specific to a single bacterial lineage, a virus, or an MGE. Additionally, EALS could be the approach of choice for projects in which only a small number of target single cells will be sequenced, making split-and-pool barcoding unnecessary. The fluorescence signal generated by PCR amplification with fluorescently labeled primers enables the identification of positive capsules, which can then be isolated using FACS. Nevertheless, conventional cytometers are unsuitable for sorting due to the size of Atrandi capsules (e.g., 70 μm diameter of capsules produced by the Atrandi Flux device). If the specialized Atrandi Styx FACS instrument is unavailable in a laboratory, a cytometer optimized for larger particles, such as the COPAS system from Union Biometrica, can be used. The EALS approach follows the workflow illustrated in Fig. 3.4b.

3.3.2.1 Cell Fixation

See the protocol in Sect. 3.3.1.1.

3.3.2.2 Microfluidic Encapsulation

A count of 62,500 cells suitable for encapsulation by the Atrandi Flux device can be achieved by precise sorting using FACS or by measuring the optical density of a bacterial sample and preparing dilutions corresponding to this cell number. The advantage of using FACS is that it allows the application of nucleic acid stains to distinguish true bacterial cells from other small particles lacking DNA. The microfluidic encapsulation protocol follows as described in Sect. 3.3.1.3.

3.3.2.3 Cell Lysis and Whole-Genome Amplification with MDA

See the protocol in Sect. 3.3.1.4.

3.3.2.4 Capsule-PCR

To identify capsules containing a target gene, capsule-PCR is performed. In this process, primers specific to the genetic element of interest are conjugated with fluorophores at their 5′ ends. The entire SPC sample obtained after whole-genome amplification is divided into individual PCR tubes for PCR setup. Each reaction contains 10.5 μL of SPCs, 12.5 μL of KAPA2G fast HG mix (REF: KK5603), 1.5 μL of nuclease-free water, and 0.5 μL of forward and reverse primers, previously diluted in water in a 1:10 ratio.

3.3.2.5 Fluorescent SPC Sorting with COPAS Flow Cytometer

Upon successful PCR amplification, the emitted fluorescence signal enables the detection and selection of positive capsules in a flow cytometer. To analyze SPC capsules, a control sample of semipermeable capsules is first used to define an initial gating region in biparametric plots (Fig. 3.5). This control sample is unprocessed and consists solely of capsules generated to provide a reference for the size

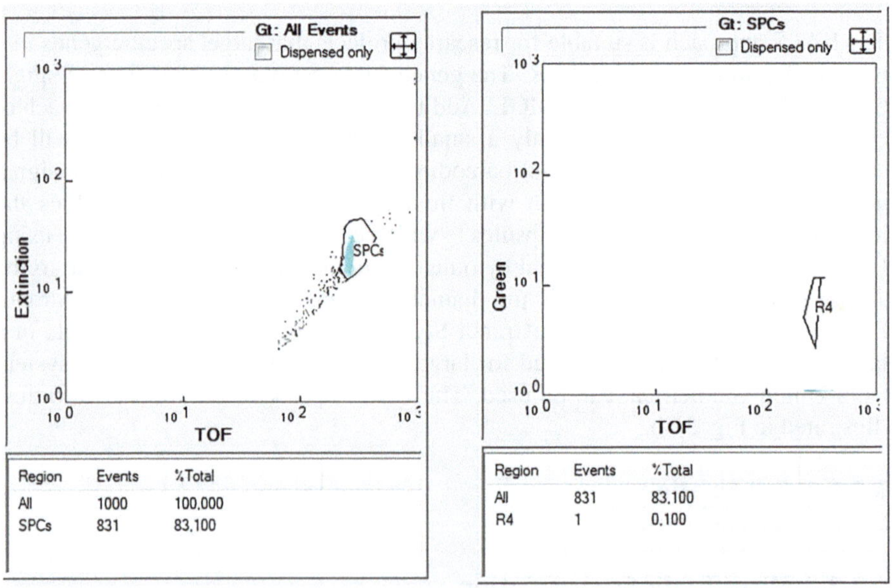

Fig. 3.5 Flow cytometry analysis of SPCs (negative control). In the gate labeled "SPCs" in the first biparametric graph, which encompasses the time-of-flight (TOF) and extinction parameters, particles corresponding to the size of the semi-permeable capsules, highlighted in blue, are concentrated. The second biparametric plot examines the fluorescence intensity emitted by the events within the "SPCs" gate of the first plot, revealing that the SPCs themselves do not emit fluorescence. Below each graph, the total number of events analyzed for each category is presented along with the percentage that each category represents of the overall total

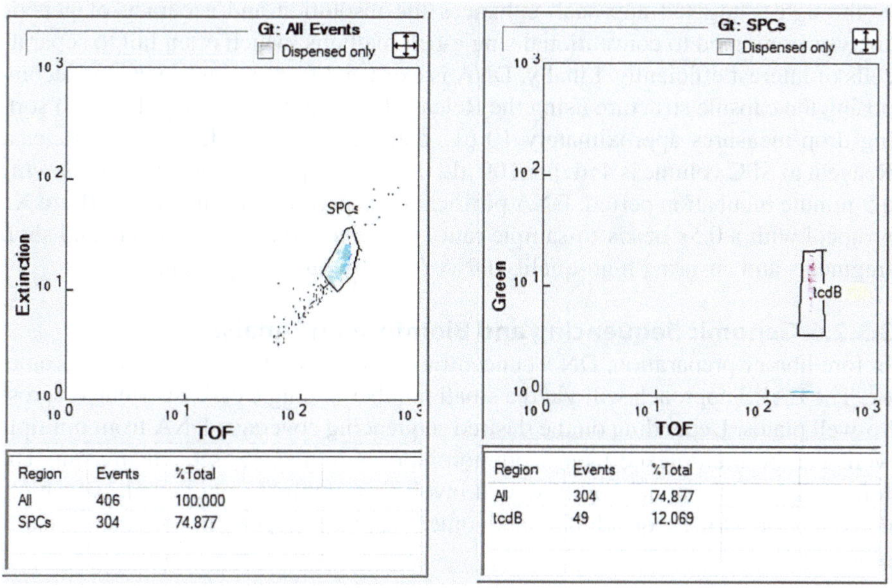

Fig. 3.6 Results of flow cytometry conducted on fluorescent semipermeable capsules (SPCs) after applying the capsule-PCR technique to amplify the tcdB gene of *Clostridioides* difficile. In the first biparametric plot, which examines Time-Of-Flight (TOF) and Extinction parameters, particles corresponding to semipermeable capsules, highlighted in blue, are concentrated within the gate labeled "SPCs." The second biparametric plot analyzes the fluorescence intensity emitted by the events identified within the "SPCs" gate from the first plot. This analysis determines the percentage of events corresponding to SPCs that underwent amplification of the tcdB gene using fluorescent primers, which are grouped within the "tcdB" gate and marked in purple. Below each plot, the total number of events analyzed for each category is provided, along with the percentage they represent of the overall events

of these structures. These plots are structured with the particle's time of flight (TOF) on the horizontal axis and extinction on the vertical axis—both parameters correlated with particle size. This process serves as a baseline to distinguish the expected size range of the capsules expected to be found before further processing the sample of interest.

The target SPC sample, which has undergone a capsule-PCR procedure using fluorophore-conjugated primers, is then analyzed in a new biparametric plot (Fig. 3.6). This plot compares the TOF parameter with fluorescence intensity detected in the green channel, allowing the quantification of fluorescent SPCs. These must contain a microbial genome that presents the target gene or other genetic elements of interest.

After fluorescence analysis, individual SPCs are sorted into a multi-well plate by instructing the cytometer to isolate fluorescent capsules identified within the second plot's gating region. Since an individual capsule contains sequences from a single cell, the analysis of the genomic content. The ability to isolate these capsules individually.

through a targeted approach enhances the resolution and accuracy of genetic analysis compared to conventional single-cell methods, which often fail to separate cells of interest efficiently. Finally, DNA is extracted from a single SPC by decomposing the capsule structure using the Release Reagent. Considering that each sorting drop measures approximately 10 μL, and the recommended ratio of Release Reagent to SPC volume is 4 μL per 100 μL, 0.4 μL is applied to each drop. Following a 5-minute incubation period, DNA purification is performed using the AMPure XP protocol with a 0.5x beads-to-sample ratio, effectively removing primers and short fragments and ensuring high-quality DNA for subsequent sequencing.

3.3.2.6 Genomic Sequencing and Bioinformatic Analysis

Before library preparation, DNA concentration is assessed using a Qubit fluorometer. The EALS approach will yield a small number of single cells distributed across 96-well plates. Depending on the desired sequencing coverage, DNA from multiple wells can be multiplexed using common indexes in a single sequencing run. The following bioinformatic workflow will involve assembly, taxonomic classification, detection of viruses or MGEs, and other analyses depending on the research objective.

3.4 Conclusions

The development and application of single-cell genomics have significantly transformed our approach to studying microbial diversity. By overcoming the limitations of traditional cultivation methods and metagenomics, single-cell genomics allows for a deeper exploration of the microbial population's genetic diversity and inherent functional capabilities. Incorporating microfluidics into the single-cell genomics workflow increases throughput, enabling the accurate reconstruction of individual genomes on a large scale, as well as facilitating the analysis of connections between bacteria and their mobile genetic elements. This enhances our understanding of microbial ecology and provides deeper insights into microbial evolution.

References

Ahsanuddin S, Afshinnekoo E, Gandara J, Hakyemezoğlu M, Bezdan D, Minot S, Greenfield N, Mason CE (2017) Assessment of REPLI-g multiple displacement whole genome amplification (WGA) techniques for metagenomic applications. J Biomol Tech 28(1):46–55. https://doi.org/10.7171/jbt.17-2801-008

Alneberg J, Karlsson CMG, Divne AM, Bergin C, Homa F, Lindh MV, Hugerth LW, Ettema TJG, Bertilsson S, Andersson AF, Pinhassi J (2018) Genomes from uncultivated prokaryotes: a comparison of metagenome-assembled and single-amplified genomes. Microbiome 6(1):173. https://doi.org/10.1186/s40168-018-0550-0

Arnau V, Ortiz-Maiques A, Valero-Tebar J, Mora-Quilis L, Kurmauskaite V, Campos Dopazo L, Domingo-Calap P, Džunková M (2025) CleanBar: a versatile demultiplexing tool for split-and-pool barcoding in single-cell omics. ISME Commun 5(1):ycaf134. https://doi.org/10.1093/ismeco/ycaf134

Berg G, Rybakova D, Fischer D, Cernava T, Vergès MC, Charles T, Chen X, Cocolin L, Eversole K, Corral GH, Kazou M, Kinkel L, Lange L, Lima N, Loy A, Macklin JA, Maguin E, Mauchline T, McClure R, Mitter B et al (2020) Microbiome definition re-visited: old concepts and new challenges. Microbiome 8(1):103. https://doi.org/10.1186/s40168-020-00875-0

Bowers RM, Kyrpides NC, Stepanauskas R, Harmon-Smith M, Doud D, Reddy TBK, Schulz F, Jarett J, Rivers AR, Eloe-Fadrosh EA, Tringe SG, Ivanova NN, Copeland A, Clum A, Becraft ED, Malmstrom RR, Birren B, Podar M, Bork P, Weinstock GM et al (2017) Minimum information about a single amplified genome (MISAG) and a metagenome-assembled genome (MIMAG) of bacteria and archaea. Nat Biotechnol 35(8):725–731. https://doi.org/10.1038/nbt.3893

Bowers RM, Gonzalez-Pena V, Wardhani K, Goudeau D, Blow MJ, Udwary D, Klein D, Vill AC, Brito IL, Woyke T, Malmstrom RR, Gawad C (2024) scMicrobe PTA: near complete genomes from single bacterial cells. ISME Commun 4(1):ycae085. https://doi.org/10.1093/ismeco/ycae085

Clingenpeel S, Schwientek P, Hugenholtz P, Woyke T (2014) Effects of sample treatments on genome recovery via single-cell genomics. ISME J 8(12):2546–2549. https://doi.org/10.1038/ismej.2014.92

Duran M, Serrano A, Nikulin A, Dauvergne J-L, Derzsi L, Palomo del Barrio E (2022) Microcapsule production by droplet microfluidics: a review from the material science approach. Mater Des 223:111230. https://doi.org/10.1016/j.matdes.2022.111230

Džunková M, La Clair JJ, Tyml T, Doud D, Schulz F, Piquer-Esteban S, Porcel Sanchis D, Osborn A, Robinson D, Louie KB, Bowen BP, Bowers RM, Lee J, Arnau V, Díaz-Villanueva W, Stepanauskas R, Gosliner T, Date SV, Northen TR, Cheng JF et al (2023) Synthase-selected sorting approach identifies a beta-lactone synthase in a nudibranch symbiotic bacterium. Microbiome 11(1):130. https://doi.org/10.1186/s40168-023-01560-8

Edwards RA, McNair K, Faust K, Raes J, Dutilh BE (2016) Computational approaches to predict bacteriophage-host relationships. FEMS Microbiol Rev 40(2):258–272. https://doi.org/10.1093/femsre/fuv048

Frost LS, Leplae R, Summers AO, Toussaint A (2005) Mobile genetic elements: the agents of open-source evolution. Nat Rev Microbiol 3(9):722–732. https://doi.org/10.1038/nrmicro1235

Garfias-Gallegos D, Zirión-Martínez C, Bustos-Díaz ED, Arellano-Fernández TV, Lovaco-Flores JA, Espinosa-Jaime A, Avelar-Rivas JA, Sélem-Mójica N (2022) Metagenomics bioinformatic pipeline. Methods Mol Biol 2512:153–179. https://doi.org/10.1007/978-1-0716-2429-6_10

Gawad C, Koh W, Quake SR (2016) Single-cell genome sequencing: current state of the science. Nat Rev Genet 17(3):175–188. https://doi.org/10.1038/nrg.2015.16

Gonzalez-Pena V, Natarajan S, Xia Y, Klein D, Carter R, Pang Y, Shaner B, Annu K, Putnam D, Chen W, Connelly J, Pruett-Miller S, Chen X, Easton J, Gawad C (2021) Accurate genomic variant detection in single cells with primary template-directed amplification. Proc Natl Acad Sci U S A 118(24):e2024176118. https://doi.org/10.1073/pnas.2024176118

Grün D, van Oudenaarden A (2015) Design and analysis of single-cell sequencing experiments. Cell 163(4):799–810. https://doi.org/10.1016/j.cell.2015.10.039

Hedlund BP, Chuvochina M, Hugenholtz P, Konstantinidis KT, Murray AE, Palmer M, Parks DH, Probst AJ, Reysenbach AL, Rodriguez-R LM, Rossello-Mora R, Sutcliffe IC, Venter SN, Whitman WB (2022) SeqCode: a nomenclatural code for prokaryotes described from sequence data. Nat Microbiol 7(10):1702–1708. https://doi.org/10.1038/s41564-022-01214-9

Hosokawa M, Nishikawa Y, Kogawa M et al (2017) Massively parallel whole genome amplification for single-cell sequencing using droplet microfluidics. Sci Rep 7:5199. https://doi.org/10.1038/s41598-017-05436-4

Ishoey T, Woyke T, Stepanauskas R, Novotny M, Lasken RS (2008) Genomic sequencing of single microbial cells from environmental samples. Curr Opin Microbiol 11(3):198–204. https://doi.org/10.1016/j.mib.2008.05.006

Kuchina A, Brettner LM, Paleologu L, Roco CM, Rosenberg AB, Carignano A, Kibler R, Hirano M, DePaolo RW, Seelig G (2021) Microbial single-cell RNA sequencing by split-pool barcoding. Science 371(6531):eaba5257. https://doi.org/10.1126/science.aba5257

Lan Y, Zhou Y, Wu M, Jia C, Zhao J (2023) Microfluidic based single cell or droplet manipulation: methods and applications. Talanta 265:124776. https://doi.org/10.1016/j.talanta.2023.124776
Lawrence D, Campbell DE, Schriefer LA, Rodgers R, Walker FC, Turkin M, Droit L, Parkes M, Handley SA, Baldridge MT (2022) Single-cell genomics for resolution of conserved bacterial genes and mobile genetic elements of the human intestinal microbiota using flow cytometry. Gut Microbes 14(1):2029673. https://doi.org/10.1080/19490976.2022.2029673
Leonaviciene G, Leonavicius K, Meskys R, Mazutis L (2020) Multi-step processing of single cells using semi-permeable capsules. Lab Chip 20(21):4052–4062. https://doi.org/10.1039/d0lc00660b
Lewis WH, Tahon G, Geesink P, Sousa DZ, Ettema TJG (2021) Innovations to culturing the uncultured microbial majority. Nat Rev Microbiol 19(4):225–240. https://doi.org/10.1038/s41579-020-00458-8
Lu N, Qiao Y, Lu Z, Tu J (2023) Chimera: the spoiler in multiple displacement amplification. Comput Struct Biotechnol J 21:1688–1696. https://doi.org/10.1016/j.csbj.2023.02.034
Luquette LJ, Miller MB, Zhou Z, Bohrson CL, Zhao Y, Jin H, Gulhan D, Ganz J, Bizzotto S, Kirkham S, Hochepied T, Libert C, Galor A, Kim J, Lodato MA, Garaycoechea JI, Gawad C, West J, Walsh CA, Park PJ (2022) Single-cell genome sequencing of human neurons identifies somatic point mutation and indel enrichment in regulatory elements. Nat Genet 54(10):1564–1571. https://doi.org/10.1038/s41588-022-01180-2
Maguire F, Jia B, Gray KL, Lau WYV, Beiko RG, Brinkman FSL (2020) Metagenome-assembled genome binning methods with short reads disproportionately fail for plasmids and genomic islands. Microbial Gen 6(10):mgen000436. https://doi.org/10.1099/mgen.0.000436
Paez-Espino D, Eloe-Fadrosh EA, Pavlopoulos GA, Thomas AD, Huntemann M, Mikhailova N, Rubin E, Ivanova NN, Kyrpides NC (2016) Uncovering Earth's virome. Nature 536(7617):425–430. https://doi.org/10.1038/nature19094
Pereira AC, Tenreiro A, Cunha MV (2022) When FLOW-FISH met FACS: combining multiparametric, dynamic approaches for microbial single-cell research in the total environment. Sci Total Environ 806(Pt 2):150682. https://doi.org/10.1016/j.scitotenv.2021.150682
Quince C, Walker AW, Simpson JT, Loman NJ, Segata N (2017) Shotgun metagenomics, from sampling to analysis. Nat Biotechnol 35(9):833–844. https://doi.org/10.1038/nbt.3935
Reichart NJ, Jay ZJ, Krukenberg V, Parker AE, Spietz RL, Hatzenpichler R (2020) Activity-based cell sorting reveals responses of uncultured archaea and bacteria to substrate amendment. ISME J 14(11):2851–2861. https://doi.org/10.1038/s41396-020-00749-1
Shibutani M, Uneyama C (2005) Methacarn fixation for genomic DNA analysis in microdissected cells. Methods Mol Biol 293:11–25. https://doi.org/10.1385/1-59259-853-6:011
Shibutani M, Uneyama C, Miyazaki K, Toyoda K, Hirose M (2000) Methacarn fixation: a novel tool for analysis of gene expressions in paraffin-embedded tissue specimens. Lab Invest 80(2):199–208. https://doi.org/10.1038/labinvest.3780023
Steen AD, Crits-Christoph A, Carini P, DeAngelis KM, Fierer N, Lloyd KG, Cameron Thrash J (2019) High proportions of bacteria and archaea across most biomes remain uncultured. ISME J 13(12):3126–3130. https://doi.org/10.1038/s41396-019-0484-y
Stepanauskas R, Fergusson EA, Brown J, Poulton NJ, Tupper B, Labonté JM, Becraft ED, Brown JM, Pachiadaki MG, Povilaitis T, Thompson BP, Mascena CJ, Bellows WK, Lubys A (2017) Author correction: improved genome recovery and integrated cell-size analyses of individual uncultured microbial cells and viral particles. Nat Commun 8(1):2134. https://doi.org/10.1038/s41467-017-02128-5
Wang Y, Navin NE (2015) Advances and applications of single-cell sequencing technologies. Mol Cell 58(4):598–609. https://doi.org/10.1016/j.molcel.2015.05.005
Wang G, Zhao L, Shi Y, Qu F, Ding Y, Liu W, Liu C, Luo G, Li M, Bai X, Li L, Ho Y-P, Yu J (2023) High-throughput generic single-entity sequencing using droplet microfluidics. bioRxiv:2023.08.15.549386. https://doi.org/10.1101/2023.08.15.549386

Weltzer ML, Wall D (2023) Social diversification driven by mobile genetic elements. Genes 14(3):648. https://doi.org/10.3390/genes14030648

Woyke T, Doud DFR, Schulz F (2017) The trajectory of microbial single-cell sequencing. Nat Methods 14(11):1045–1054. https://doi.org/10.1038/nmeth.4469

Yang C, Chowdhury D, Zhang Z, Cheung WK, Lu A, Bian Z, Zhang L (2021) A review of computational tools for generating metagenome-assembled genomes from metagenomic sequencing data. Comput Struct Biotechnol J 19:6301–6314. https://doi.org/10.1016/j.csbj.2021.11.028

FISH–FACS Enabled Targeted Recovery of Genomes from Uncultivated Environmental Microbial Populations

4

Anissa Grieb, Monike Oggerin, Janey Lee, Danielle Goudeau, Rex R. Malmstrom, Tanja Woyke, and Bernhard M. Fuchs

Abstract

As most microbial diversity on Earth remains uncultivated, metagenomics and single-cell genomics have been established as crucial approaches for studying microbial assemblages. While these methodologies provide invaluable insights into the genetic makeup of microbial community members, the genomes recovered most often represent the abundant microorganisms or cells randomly captured. This limits the targeted recovery of genetic material from specific taxa of interest. We here present a comprehensive protocol for the targeted recovery of genomes from uncultivated environmental microorganisms by combining fluorescence in situ hybridization (FISH) with cell sorting. The protocol consists of (i) fixation of cells; (ii) labeling with hybridization chain reaction-FISH; (iii) fluorescence-activated cell sorting (FACS) of labeled populations. The genomic DNA of sorted cells can then be amplified and sequenced using previously established protocols. This procedure can be performed with standard molecular biology and fluorescence imaging equipment, including a FACS machine, and completed within 24 h. This protocol can be adapted for any bacterial or archaeal taxonomic group of interest with available 16S rRNA sequence information.

A. Grieb
Max Planck Institute for Marine Microbiology, Bremen, Germany

DVGW-Research Centre TUHH, Technical University Hamburg, Hamburg, Germany

M. Oggerin · B. M. Fuchs (✉)
Max Planck Institute for Marine Microbiology, Bremen, Germany
e-mail: bfuchs@mpi-bremen.de

J. Lee · D. Goudeau · R. R. Malmstrom · T. Woyke (✉)
Department of Energy Joint Genome Institute, Berkeley, CA, USA
e-mail: twoyke@lbl.gov

C. Rinke, M. Džunková (eds.), *Single-Cell Omics in Microbiome Research*, The Microbiomes of Humans, Animals, Plants, and the Environment 7,
https://doi.org/10.1007/978-3-032-07527-7_4

Keywords

Targeted cell sorting · Whole genome amplification · Bioinformatics · Fluorescence in situ hybridization · FISH · Mini-metagenomes · Fixation

4.1 Introduction

The continued advances in sequencing methods, bioinformatic data assembly, and binning approaches have enabled the characterization of vast swaths of microbial diversity, including species not cultivated in the laboratory. Shotgun metagenomics, in particular, reveals invaluable insights into the genetic makeup of bacteria and archaea across various ecosystems (Albertsen et al. 2013; Anantharaman et al. 2016). It is complemented by single-cell sequencing, which, despite being lower in throughput and hampered by partial genome completeness (Rinke et al. 2013; Woyke et al. 2017), provides links to phages and other mobile genetic elements and reveals population heterogeneity at the single-cell level (Kashtan et al. 2014; Engel et al. 2014; Bowers et al. 2022). However, both approaches are untargeted, capturing the genomes of abundant species more often than rare species. This hampers the recovery of high-quality genomes from members of the rare biosphere, which, despite their low abundance, have been shown to play key ecosystem roles (Hausmann et al. 2016; Wang et al. 2017; Jousset et al. 2017).

4.1.1 Development of the Protocol

Recent studies have shown that reducing diversity within a metagenome can facilitate the assembly of lower-abundance species. These mini-metagenomes have been created through random subsampling of the total population (Ji et al. 2017; Yu et al. 2017; Schulz et al. 2018; Berghuis et al. 2019) or separation based on functions (Jing et al. 2018; Doud et al. 2020). Subsampling based on taxonomic affiliation, using fluorescence in situ hybridization (FISH) and cell sorting, has been applied in several studies (Wallner et al. 1997; Sekar et al. 2004; Kalyuzhnaya et al. 2006; Miyauchi et al. 2007; Yilmaz et al. 2010). However, these studies have sequenced only specific genes, such as the 16S rRNA gene, from the sorted cells. Thus far, genome sequencing of a specific taxonomic group of interest, subsampled based on FISH, has not led to high-quality assemblies (Podar et al. 2007). Therefore, we developed a FISH–FACS approach broadly applicable for targeting and capturing whole genomes from specific uncultivated taxa of interest.

In principle, FISH is used to fluorescently label microbial groups of interest for flow cytometric cell sorting and subsequent whole genome sequencing. In a recent study, we labeled and sorted an uncultivated *Bacteroidota* genus from a planktonic seawater sample from the North Sea, which yielded good-quality (>80% completeness, <5% contamination) metagenome assembled genomes (MAGs) from the sorted mini-metagenomes (Grieb et al. 2020).

For the detection of FISH-hybridized cells via flow cytometry, a bright FISH signal is critical. The short observation times combined with the lower sensitivity of flow cytometers, compared to modern epifluorescence microscopes, make strong FISH signals essential for effective cell sorting. While strong FISH signals have been achieved from nutrient-rich environments like wastewater, soil, oral mouth cavity, and gut (Campbell et al. 2013; Bruder et al. 2016; Hausmann et al. 2016; Pratscher et al. 2018; Dam et al. 2020), this has—to date—not yet been demonstrated for dilute, nutrient-poor environmental samples, where cells often have small sizes and low ribosome contents. Strong fluorescence signals can be provided with enzyme-labeled oligonucleotide probes through catalyzed–reporter–deposition–FISH (CARD–FISH) (Pernthaler et al. 2002). However, CARD–FISH typically requires fixation with formaldehyde to maintain cell integrity. Formaldehyde fixation, however, inhibits genome amplification by Multiple Displacement Amplification (MDA) (Clingenpeel et al. 2014), the predominant whole-genome amplification method used for single cells. Furthermore, CARD–FISH relies on free radical reactions during the signal amplification that damage DNA (Pernthaler et al. 2002).

We previously demonstrated strong fluorescent labeling of bacterial isolates and environmental microbiome cells using ethanol fixation and Hybridization Chain Reaction (HCR)–FISH. This approach is an alternative to CARD–FISH, which does not utilize radical reactions and thus preserves the integrity of cellular DNA for downstream DNA amplification (Grieb et al. 2020). When targeting specific cells from nutrient-deprived environments using FACS, ethanol-fixation combined with HCR–FISH is recommended.

4.1.2 Applications of the Method

The main application of the pipeline is the retrieval of genomes from uncultivated bacterial or archaeal taxa identified in an environmental sample, particularly through methods such as 16S rRNA gene amplicon sequencing. This targeted approach is especially beneficial for enriching and recovering rare biosphere taxa found in complex communities, which are often challenging to capture using untargeted, cultivation-independent methods. The pipeline was developed for dilute planktonic microorganisms collected during marine research cruises on polycarbonate filters and has yet to be tested for other environments.

4.1.3 Advantages & Limitations

The primary advantage of the described protocol is its ability to efficiently access genomes from uncultivated taxa of interest by directly capturing them from an environmental sample, facilitating their genomic characterization. This process relies on a priori knowledge of a specific bacterial or archaeal group, typically based on 16S rRNA amplicon sequence information. Some advantages over untargeted cultivation-independent genomics approaches include (1) concentrating efforts and sequencing resources on only the microorganism(s) of interest; (2) enriching rare taxa that can be overwhelmed by abundant groups; (3) linking 16S rRNA information directly to the recovered genomes. For traditional single-cell sequencing, the cells are captured randomly (Rinke et al. 2014). In shotgun metagenomes, DNA from all community members is sequenced, and as a result, genomes from rare biosphere members may be missed, depending on the sequencing depth for shotgun metagenomics (Lynch and Neufeld 2015). In addition, binning tools often fail to accurately assemble and assign 16S rRNA gene sequences to MAGs due to the high degree of conservation of the 16S rRNA gene (Chen et al. 2020). By using a 16S rRNA-based probe for targeted subsampling and sequencing, this approach establishes a direct experimental link between the genomic information and the taxonomic classification based on the 16S rRNA information.

One limitation of the HCR–FISH–FACS pipeline is the potential challenge of designing specific probes for the intended target group. As described by Amann and Fuchs (2008), careful, in-depth phylogenetic reconstruction is crucial to accurately define the target group and to develop novel probes. Existing FISH probes need to be checked against an up-to-date database to verify their specificity. Even then, this specificity is limited to the currently known species and cannot account for the yet unknown microorganisms (Amann and Fuchs 2008). Additionally, some species with potentially very different physiologies may not be delineated on the 16S rRNA level yet are targeted by the same FISH probe (Amann and Fuchs 2008). Along these lines, the protocol requires a priori knowledge of the 16S rRNA sequence of microorganisms of interest and their presence in a given environmental sample. This requirement limits its application for the de novo analysis of unknown taxa and environments. Lastly, despite signal amplification provided by HCR–FISH relative to standard FISH, the fluorescence intensity is still influenced by the number of ribosomes, which could hamper the detection of cells with a low number of ribosomes due to the low FISH signal.

4.1.4 Experimental Design

The HCR–FISH–FACS protocol is divided into three experimental parts: (i) cell fixation, (ii) fluorescence in situ hybridization to fluorescently label the microbial group of interest, and (iii) fluorescence-activated cell sorting to separate the labeled cells from the unlabeled cells. Genomic DNA of sorted cells can then be amplified using multiple displacement amplification per previously established protocols (Rinke et al. 2014; Rinke 2018), followed by genome sequencing. An overview of the experimental workflow is depicted in Fig. 4.1.

4.1.4.1 Cell Fixation

All HCR–FISH steps described in this protocol are performed on cells collected on polycarbonate filters. Using a vacuum pump, a water sample is filtered through larger pore size filters to remove aggregates that could clog the flow cytometer

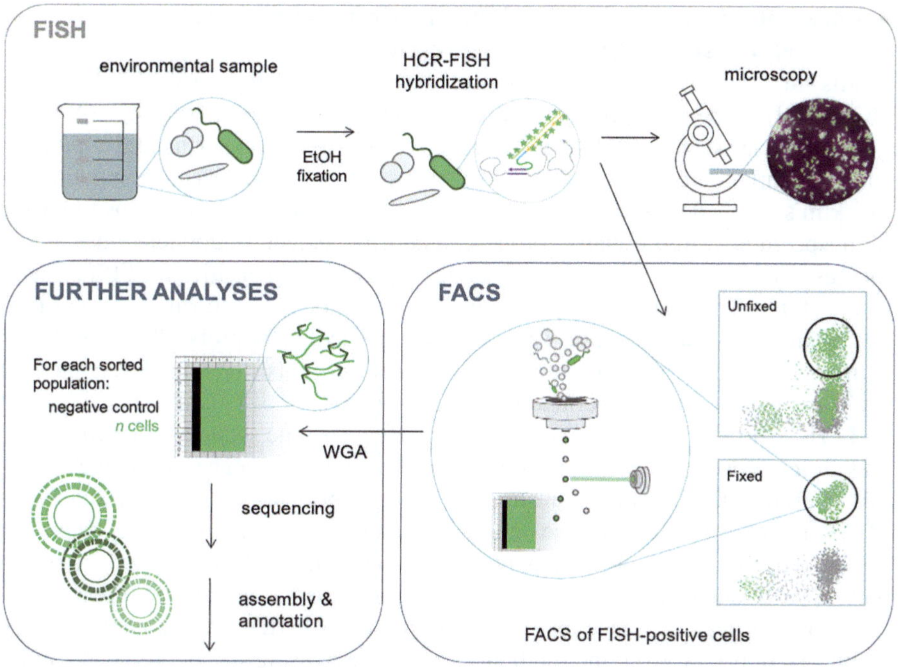

Fig. 4.1 Experimental workflow. FISH—water samples are filtered and fixed with ethanol before the cells are taxon-specifically labeled using HCR–FISH. Successful hybridization can be verified by microscopy. FACS—hybridized cells are removed from the filter and sorted based on their fluorescent HCR–FISH signal. Ethanol-fixed cells show a brighter HCR–FISH signal than unfixed cells (Grieb et al. 2020). Further analyses—The FISH–FACS pipeline has been optimized for subsequent Whole Genome Amplification (WGA) and sequencing. While we recommend sorting at least 500 cells, as based on previous work (Grieb et al. 2020), the number of cells sorted can be adjusted depending on the specifics of a given sample and the experimental goal

nozzle. The pore size is thereby dependent on the size fraction of interest. Using a 10 μm pore-size prefilter provides a good compromise between removing large particles and retaining cells, but might be increased to up to 40 μm pore-size for a 70 μm diameter FACS nozzle. The final filtration step uses a 0.2 μm pore size to fix the cells on the filter surface.

A 15 min incubation of the filter containing the cells in 96% ethanol fixes the cells by dehydration (Step 1 (B)). Cells from culture are treated with 70% ethanol before filtration and storage (Step 1 (A)). Ethanol fixation has been shown to provide the best signal intensity without impairing the DNA quality for sequencing (Grieb et al. 2020). All further HCR–FISH hybridization and amplification steps are performed with these cells on the filter. Fixed filters can be stored at −20 °C.

4.1.4.2 Labeling of Cells with Hybridization Chain Reaction—Fluorescence In Situ Hybridization

In this step, the targeted cells are fluorescently labeled using a 16S rRNA-specific HCR–FISH probe. This probe can be designed with tools like ARB using 16S rRNA gene databases such as SILVA (www.arb-silva.de). Testing experimental conditions such as temperature and formamide concentrations, and considering competitors to block mismatched close relatives (Manz et al. 1992) are necessary steps to ensure high specificity. Additionally, probe-neighboring helper oligonucleotides might increase hybridization efficiency to provide high signal intensities for FISH (Fuchs et al. 2000). The specificity of existing probes should be affirmed by checking against an up-to-date database. The described protocol is for probes that need a 35% formamide concentration in the hybridization buffer. This step must be adapted for probes requiring different formamide concentrations.

In HCR–FISH, the probes consist of the target specific sequence (usually 18 nt) and an initiator sequence (26 nt) linked by an adaptor sequence (Yamaguchi et al. 2015). During the hybridization step, the probe hybridizes to ribosomal sequences of the target cells, leaving the initiator sequence overhanging. After removing unbound probes in a washing step, two fluorescently labeled hairpin monomers (H1 and H2) are added. H1 can attach to the initiator sequence and unfolds upon attachment, presenting a sequence for H2 to bind and unfold. This process establishes the chain reaction that gives the method its name, increasing the fluorescence signal with each attached hairpin monomer.

The HCR–FISH protocol was modified from Yamaguchi et al. (2015) with substantial changes encompassing a hybridization buffer that was adapted from the geneFISH protocol (Barrero-Canosa et al. 2017) and the introduction of a heating step before hybridization. The geneFISH protocol also inspired HCR–FISH but with reduced temperature (65 °C instead of 85 °C) that helps unfold secondary structures of the probe and target rRNA.

The use of two additional probes as negative and positive controls should be considered to facilitate detection in the following cell sorting step. As a negative control, a nonbinding probe such as the NON338-probe (Wallner et al. 1993) can be used. Alternatively, no probe is added to the hybridization step. As a positive control, a probe for a higher taxonomic rank can be used.

HCR–FISH hairpin molecules can be labeled with different fluorochromes to match the detection properties of the used flow cytometer. The probe fluorochrome should be bright and easily detectable. Counterstains such as DAPI or SYTO dyes to stain all nontarget cells should be adapted accordingly and spectrally separated from the probe fluorochrome, ideally excited by a different laser. Microscopy of FISH-treated cells confirms successful hybridization (Fig. 4.2) and provides the relative abundance of the targeted group in the sample.

4.1.4.3 Fluorescence-Activated Cell Sorting (FACS)

To analyze the cells by flow cytometry, they are removed from the filter. This can be done in a buffer containing 0.05% Tween80 using vigorous vortexing or sonication (steps 24 (A) and (B)). Previous studies (Sekar et al. 2004) have shown that cell loss

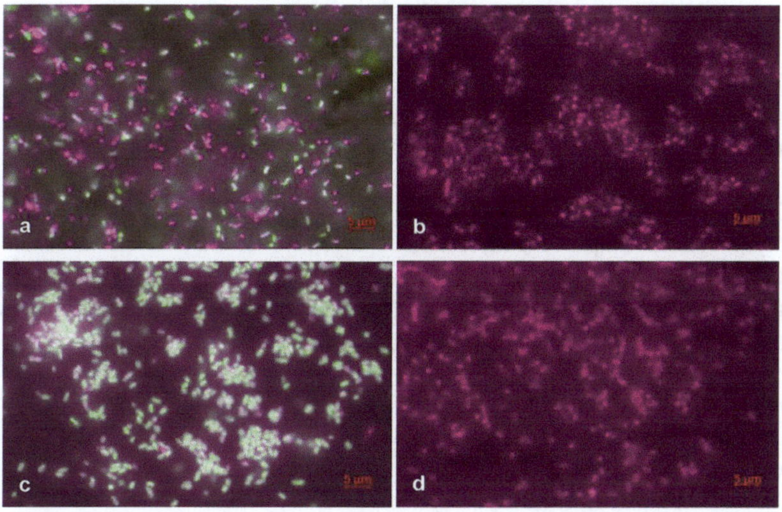

Fig. 4.2 Images show *Escherichia coli* cells that have undergone HCR–FISH labeling (green signal) in both unfixed (**a**) and 70% ethanol fixed cultures (**b**) and unfixed cells (**c**). HCR–FISH reactions excluding the initiator probe were also performed as negative controls of fixed (**d**). Control samples (no initiator or a nonbinding initiator) should be without an HCR–FISH signal and show only DNA-staining (cyan signal). Ethanol-fixed cells show more hybridized intact cells than unfixed cells

is mostly below 10%. This protocol was developed for planktonic water samples. Adaptations for samples where bacteria are attached to particles or are part of a biofilm are likely necessary in order to sort them by flow cytometry. Methods for this purpose have been described previously (Rinke et al. 2014) and include vortexing, sonication, filtration, and centrifugation steps.

The described protocol creates population pools of 500 cells, a number we recommend based on previous work (Grieb et al. 2020); however, the exact number can be customized for a given experiment. We suggest collecting multiple technical replicates of sorted cells on the same plate. In addition, depending on the study's goal, it may be desirable to sort an equivalent number of nontarget cells, i.e., cells stained with a generic DNA stain but not labeled with FISH probes, to serve as a control and reference point.

When targeting low-abundance populations, the scattered positive signals may not be easily discernible as a population. Using the positive control described above can be helpful as a larger population of a higher taxonomic rank is easier to detect and indicates where to expect the low-abundance targeted population. It is highly recommended to check the sorted cells by fluorescence microscopy to confirm that the sorting parameters are correctly set and that only hybridized cells are sorted.

4.2 Materials

4.2.1 Reagents

Sample Preparation and Preservation

- 10X phosphate buffer saline (PBS).
- Ethyl alcohol, 200 proof (Pharmco).
- Ultrapure water, such as Milli-Q water (Millipore) or molecular biology grade water (e.g., Fisher), referred to as MQ water in the remainder of this protocol.

Probe Hybridization

- Dextran sulfate sodium salt (Sigma-Aldrich).
- 20X SSC buffer (ThermoFisher Scientific).
- 0.5 M EDTA, pH 8.0 (VWR).
- Formamide (Sigma-Aldrich).
- Sheared salmon sperm DNA, 10 mg ml^{-1} (Ambion).
- Yeast RNA, 10 mg/ml (Ambion).
- Blocking reagent (Roche).
- Maleic acid (Sigma-Aldrich).
- 10% SDS (Ambion).
- 5 M NaCl (Sigma-Aldrich).
- 1 M Tris–HCl, pH 8.0 (ThermoFisher Scientific).
- Initiator probe with adaptor (5′-CCG AAT ACA AAG CAT CAA CGA CTA GA-3′), linker (5′-AAA AA-3′) (Yamaguchi et al. 2015), and probe sequence (depends on target group) (Integrated DNA Technologies (IDT) or Biomers.net).

Washing

- 1 M Tris–HCl, pH 8.0 (ThermoFisher Scientific).
- 0.5 M EDTA, pH 8.0 (VWR).
- 5 M NaCl (Sigma-Aldrich).
- 10% SDS (Ambion).

Hybridization Chain Reaction Amplification

- Sodium phosphate dibasic solution (Na_2HPO_4) (Sigma-Aldrich).
- 5 M NaCl (Sigma-Aldrich).
- Blocking reagent (Roche).
- Dextran sulfate sodium salt (Sigma-Aldrich).
- 10% SDS (Ambion).

- *H1*-4xAtto488-labeled hairpin amplifier probe (5′-TCT AGT CGT T(G)A TGC TTT (G)TA TTC GGC GAC A(G)A TAA CCG AAT ACA AA(G) CAT C-3′) (Yamaguchi et al. 2015) (Biomers.net); fluorescence labels are linked to the (G).
- *H2*-4xAtto488-labeled hairpin amplifier probe (5′-CCG AAT ACA AA(G) CAT CAA C(G)A CTA GAG ATG CTT T(G)T ATT CG(G) TTA TCT GTC G-3′) (Yamaguchi et al. 2015) (Biomers.net); fluorescence labels are linked to the (G).

Microscopy and Imaging

- 4′,6-Diamidino-2-phenylindole (DAPI), 5 mg (Sigma-Aldrich).
- Citifluor AF1 mountant solution (AF1–25, Electron Microscopy Sciences, USA).
- Vectashield Antifade Mounting Medium (H-1000-10, Vector Laboratories, USA).

Cell Resuspension From Filter for Flow Cytometry

- Tween 80 (Sigma-Aldrich).
- 5 M NaCl (Sigma-Aldrich).

4.2.2 Equipment

Sample Preparation and Preservation

- Filter with 40 μm pore size (e.g., BD Falcon Cell Strainer 40-μm; BD Biosciences).
- 0.2 μm pore-size polycarbonate filter, 25 mm (Isopore).
- 25 mm vacuum filtration assembly with 125-ml flask (Sigma–Aldrich).
- 2 ml microcentrifuge tubes (e.g., Eppendorf Safe-Lock Tubes 2.0 ml, colorless; Eppendorf).

Probe Hybridization

- Hybridization chamber, such as a glass or plastic container with a volume of about 500 ml with an airtight locking lid (e.g., Glasslock).
- Standard oven or incubator, such as VWR Horizontal Air Flow Oven 1330FMS (VWR).
- Parafilm M, Benis (VWR).
- Duraseal sealing film (Sigma-Aldrich).
- Falcon disposable petri dishes, 100 × 15 mm (Corning).
- Kimtech Science Kimwipes, 11 × 21 cm (Kimberly–Clark).

Washing

- 50-ml polypropylene tubes (ThermoFisher Scientific).
- Flat, round-tip forceps (Fisher).

H1/H2 Hairpin Preparation

- 0.2-ml PCR tubes (USA Scientific).
- Standard thermocycler, such as Applied Biosystems ProFlex PCR System (ThermoFisher Scientific).

Microscopy and Imaging

- Frosted micro slides, 75 × 25 mm (VWR).
- Coverslips, plastic, 22 mm (VWR).

Cell Resuspension from the Filter for FACS

- Probe sonicator, such as Bandelin Sonopuls HD70 (Bandelin).

Preparation of FACS for Clean Sorting

- Prepare 4 L of 1X PBS in two 2-L quartz flasks and mix with magnetic stir bars and plates within the clean hood. Position the empty sheath fluid tank and inverted lid inside the hood so that UV will shine on the inner surfaces. Close the hood and start the overnight UV exposure. We suggest dedicating one tank to clean sheath fluid. After UV exposure, carefully add sheath fluid to the tank inside the clean hood. Save at least 10 ml of clean sheath fluid for later use while sorting.

  ▲ **Critical Step** A clean FACS instrument and sheath fluid are essential for successful sorting and DNA amplification. Follow the manufacturer's instructions for instrument cleaning. If compatible with your instrument and filters, we recommend cleaning with 10% bleach (Rinke et al. 2014).

Cell Sorting by Flow Cytometry

- Microwell plates.

4.2.3 Reagent Setup

Sample Preparation and Preservation

70% Ethanol—Prepare fresh for every sample using 200-proof ethyl alcohol and molecular biology-grade water.

Probe Hybridization

Initiator Probe—In HCR–FISH, the sequence of FISH probes targeting the ribosomal RNA (usually oligomers of 15–30 nucleotides length) are extended by an adapter and initiator sequence at the 5′-end of the FISH probe. As rRNA binding sequences, published FISH probes (e.g., probeBase (Greuter et al. 2016)) or newly designed probes can be used. The latter can be created using the PROBE_DESIGN tool of the ARB software package (Ludwig et al. 2004) and a 16S rRNA database such as the SILVA database (Quast et al. 2013; Yarza et al. 2014; Glöckner et al. 2017). Through websites of Integrated DNA Technologies (IDT) or Biomers.net, 20 nmol of the probe of choice is ordered. Select parameters of purification, like HPLC purification. Resuspend the delivered oligonucleotide in MQ water to 10 µM and store at −20 °C.

Maleic Acid Buffer—Dissolve 0.46 g maleic acid (final concentration of 0.1 M) in 30 ml MQ water. Add 1.2 ml 5 M NaCl (150 mM) and mix well. Adjust the solution with 6 M NaOH to a final pH of 7.5. Bring up the maleic acid buffer to 40 ml with MQ water.

10X Blocking Reagent—To make the 10X blocking reagent, dissolve 2 g blocking reagent (Roche) in 20 ml of maleic acid buffer for a final concentration of 10% (w/v). Sterilize stock solution by autoclaving. Alternatively, add stock solution to a glass flask and cover with Duraseal sealing film. Sterilize by microwaving and stirring intermittently. Store stock solution at 4 °C.

Gene-FISH Buffer with 35% Formamide—For 40 ml of gene-FISH buffer, dissolve 8 g dextran sulfate (final concentration of 20%) in 8.2 ml MQ water. Add 10 ml 20X SSC (5X) and 1.6 ml 0.5 M EDTA (20 mM). Heat to 60 °C and mix well until dissolved. In a fume hood, combine 14 ml formamide (35%), 1 ml 10 mg ml^{-1} sheared salmon sperm DNA (0.25 mg ml^{-1}), 1 ml 10 mg ml^{-1} yeast RNA (0.25 mg ml^{-1}), and 4 ml 10X blocking reagent (1%). Add the fume hood mixture to the heated dextran sulfate solution and mix well. Add last 200 µl 20% SDS (0.1%). Mix well and pass through a 0.2-µm pore-size filter. Store at −20 °C.

! Caution Buffers that contain formamide should be prepared under the fume hood, as it is a teratogenic compound and is volatile above 40 °C.

▲Critical Step With high concentrations of formamide, the dextran sulfate might not dissolve without adding the formamide. Everything should be completely dissolved before filtration and should only be slightly milky.

Humidity Chamber Solution—Prepare a salt/formamide solution of 0.9 M NaCl and formamide matching the concentration needed for the probe. For the humidity chamber described in this protocol, approximately 5 ml is required.

Washing (1)

Wash Buffer—Combine the following: 47.7 ml MQ water, 0.5 ml 0.5 M EDTA (final concentration of 5 mM), 1 ml 1 M Tris–HCl, pH 8.0 (20 mM), and 800 µl 5 M NaCl (80 mM). Add the last 50 µl 10% SDS (0.01%) and mix well. Make fresh with every incubation and preheat before use to 48 °C.

The concentration of NaCl is dependent on the used formamide concentration (https://www.arb-silva.de/fileadmin/graphics_fish/SILVA_FISH_protocols_mono_101025_V2_2.pdf).

Hybridization Chain Reaction Amplification

HCR Buffer 2—For 20 ml of HCR buffer 2, dissolve 2 g dextran sulfate (to a final concentration of 10%) in 12.4 ml MQ water. Add 2 ml 0.5 M Na_2HPO_4 (50 mM), 3.6 ml 5 M NaCl (0.9 M), and 2 ml 10% blocking reagent (1%). Heat to 60 °C and dissolve fully before adding 20 µl 10% SDS (0.01%). Mix well and pass through a 0.2-µm pore-size filter. Aliquot into 1.5 ml Eppendorf tubes and store at −20 °C.

Microscopy and Imaging

DAPI Working Solution—Using MQ water, make a 3.6 mM (1 mg ml^{-1}) DAPI stock solution. Aliquot and store at −20 °C. For a 3.6 µM (1 µg ml^{-1}) DAPI working solution, dilute 1:1000 in MQ water.

Mounting Medium—Mix Citifluor AF1 mountant solution and Vectashield Antifade Mounting Medium in a 9:1 ratio. Store at −20 °C.

Cell Resuspension from Filter for Flow Cytometry

Cell Resuspension Buffer—Prepare MQ water solution with 150 mM NaCl and 0.05% Tween 80, as described in Sekar et al. (2004).

4.2.4 Procedure

Sample Collection, Preparation, and Preservation •Timing up to 2 h (depending on filtration speed).

1. Sample processing procedures and their time requirements vary depending on the type of sample—bacterial cell cultures or water samples:

   (A) Bacterial Cultures Preserved by Ethanol Fixation

      (i) Grow cultures on a shaker in liquid media to an OD_{600} of ~0.1 at appropriate growth conditions.
      (ii) Pellet cells at 10,000× *g* for 10 min. Decant supernatant and resuspend pellet in freshly made 70% ethanol.
      (iii) Incubate at room temperature for 1 h and then transfer to −20 °C for long-term storage.

      ■ **Pause Point** Ethanol-fixed cells can be stored at −20 °C for several months.
      (iv) Capture the sample on a 0.2 μm polycarbonate filter.

   (B) Ocean or Freshwater Sample Preserved by Ethanol Fixation

      (v) Pre-filter the sample (100–500 ml containing app. 10^8–10^9 cells) through a 40-μm cell strainer to remove large particulate matter and avoid clogging during the next filtration step. An additional prefiltration at 5 μm or 3 μm is recommended to obtain constrained flow cytometry signals.
      (vi) Capture the sample on a 0.2-μm polycarbonate filter.
      (vii) Place the filter into a Petri dish with 96% ethanol.
      (viii) Incubate at room temperature for 15 min, let dry, and transfer to −20 °C or −80 °C for long-term storage.

      ■ **Pause Point** Ethanol-fixed filters can be stored at −20 °C for several months.

Probe Hybridization •Timing 3 h

The HCR–FISH hybridization and amplification steps are described for a humidity chamber (appr. 500 ml) with Petri dishes (Fig. 4.3). The Petri dishes can be replaced by microslides or another level surface that fits into a humidity chamber.

2. Bring the sample to room temperature. Cut a large piece off the filter for the target probe and smaller pieces for the negative control and the optional positive control.

   ▲ **Critical Step** Filters hybridized without initiator probes or with a nonbinding probe will serve as negative controls. Filters hybridized with a probe for a higher taxonomic rank than the target probe can serve as a positive control.

   ▲ **Critical Step** Make sure to label filters with a pencil at the edge to mark the side containing cells.
3. Prepare one Petri dish per sample and attach a piece of Parafilm to each. The plastic can be labeled accordingly "with initiator probe", "without initiator probe/with nontarget probe", and "positive control".
4. Prepare a humidity chamber using an air-tight glass or plastic container. Place a paper tissue in an empty Petri dish and soak it with the humidity chamber solu-

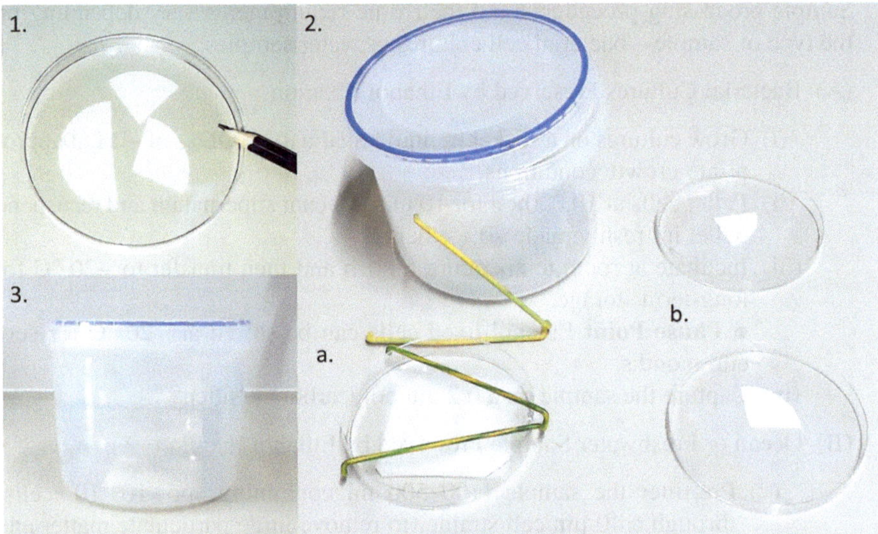

Fig. 4.3 Setup of humidity chamber. (**1**) Filters are labeled with pencil and cut into required sizes. (**2**) The humidity chamber contains (**a**) Petri dish with salt/formamide mixture (hybridization, step 4) or water (amplification, step 11) and (**b**) Petri dishes with parafilm and hybridization buffer (step 6) or amplification buffer (step 14) with filter pieces (cell side facing down). (**3**) Petri dishes are stacked in a humidity chamber, e.g., using strips of insulated electrical wire as spacers (examples shown in 2) between the dishes

tion (as prepared in REAGENT SETUP). Place the formamide/salt solution at the bottom and secure the lid to the humidity chamber.

▲**Critical Step** The humidity chamber prevents evaporation of the geneFISH buffer droplets during long incubation times.

▲**Critical Step** Soaking the tissue on the bottom of the glass container with a salt solution/formamide mix prevents hygroscopic water accumulation on the FISH preparation while hybridizing.

▲**Critical Step** It is helpful to have the humidity chamber prepared and pre-incubated at 65 °C before preparing filters for hybridization.

▲**Critical Step** If microslides are used that are directly placed on soaked tissues, they should NOT contain any pooling water that could interfere with the integrity of the geneFISH buffer droplets. Take extra care with the Kimwipes inside the Petri dishes containing the microslides. They should be damp but not dripping with water.

! Caution Solutions that contain formamide should be prepared under the fume hood, as it is a teratogenic compound and is volatile above 40 °C.

5. Thaw geneFISH buffer (as prepared in REAGENT SETUP). For each FISH reaction, add 10 μl of 10 μM probe working solution to 90 μl geneFISH buffer for a final probe concentration of 1 μM in a volume of 100 μl. For negative controls, add 10 μl MQ water or 10 μl of 10 μM nontarget probe working solution to 90 μl geneFISH buffer.

6. For each reaction, pipette 100 µl of geneFISH buffer with probe onto the Parafilm-covered slide. Cover each droplet with the filter pieces containing the sample, with the microbe side facing down into the buffer. Carefully stack Petri dishes with filters in the humidity chamber, using a separator between the Petri dishes. Secure the lid to the humidity chamber and ensure the lid fits tightly.

   ▲**Critical Step** Due to the Parafilm, the buffer should bead together in a large droplet. The filter piece should lay on top of the droplet's surface, not immersed within it.

   ▲**Critical Step** Mark the side of the filter containing the microbial population before transferring the filter to the hybridization buffer.

   ! Caution Solutions that contain formamide should be prepared under the fume hood, as it is a teratogenic compound and is volatile above 40 °C.
7. Incubate the humidity chamber in an oven at 65 °C for 30 min for denaturation.

   ▲**Critical Step** Before moving to the next step, examine the filters and the buffer droplets for any sign of drying or evaporation through the glass. Ensure no drying occurs and the entire filter stays in contact with the buffer droplet.
8. After the denaturation step, incubate the humidity chamber at 46 °C for 2 h.
9. Preheat wash buffer (as prepared in REAGENT SETUP) at 48 °C in a water bath.

Washing (1) •Timing 40 min

10. Remove the Petri dishes from the humidity chamber. Place each filter in a 50 ml conical tube filled with 20 ml of prewarmed wash buffer.

    ▲**Critical Step** Before transferring filters to the wash buffer, mark the side of the filter containing the microbial population.
11. Incubate filters in wash buffer at 48 °C for 30 min.

    ▲**Critical Step** Prepare the humidity chamber for future steps by wetting Kimwipes, this time with water, and resetting the chamber to 37 °C.

    ▲**Critical Step** Continue with steps 12–14 during this washing step to avoid delays in removing filters from the wash buffer in step 15.

Hybridization Chain Reaction Amplification •Timing 2.5 h

12. For each FISH reaction, prepare 50 µl of H1 and 50 µl of H2 in separate 0.2 ml PCR tubes, as follows: Add 25 µl HCR Buffer 2 (as prepared in REAGENT SETUP) to 25 µl of 10 µM H1-4xAtto488 (for a final concentration of 5 µM) in a 0.2 ml PCR tube. Repeat with H2 hairpin: add 25 µl HCR Buffer 2 to 25 µl of 10 µM H2-4xAtto488 (5 µM) to a separate 0.2 ml tube.
13. Incubate tubes in a thermocycler with the following cycling program: 95 °C for 90 s, 25 °C for 1 min, then keep at room temperature in the dark.
14. Prepare and label clean Petri dishes with Parafilm for each filter. Combine H1 and H2 hairpin mixtures shortly before usage for a total of 100 µl per FISH reaction. Pipette the 100 µl H1 + H2 mixture onto each slide.

▲**Critical Step** Ensure that H1 and H2 are mixed 5 min before usage.

15. Remove each filter from the wash buffer and briefly blot onto a dry, clean Kimwipe to remove excess wash buffer without letting filters completely dry. Place the filter onto the H1 + H2 droplet, keeping the microbe side facing down into the mixture.

    ▲**Critical Step** Remove filters from the wash buffer with care. Use round-tipped forceps to avoid tearing of filters and retrieve filters gently without wrinkling. Wrinkling or folding of the filters results in uneven hybridization and uneven fields of view for microscopy.

    ▲**Critical Step** Do not allow the filter to dry completely before adding it to the H1 + H2 buffer.

    ▲**Critical Step** Due to the Parafilm, the buffer should bead together in a large droplet. The filter piece should lie on top of the droplet's surface, not be immersed within it.

16. Carefully stack Petri dishes with H1 + H2 droplets into the prepared humidity chamber, using spacers, and close the lid tightly. Incubate the humidity chamber at 37 °C for 2 h in the dark.

    ▲**Critical Step** It is essential to minimize exposure of H1 and H2 to light.

Washing (2) •Timing 20 min

17. For each filter, prepare a series of three wash buffers in individual 50-ml conical tubes and incubate for the following length of time:

    (i) 20 ml 1X PBS at 4 °C for 5 min.
    (ii) 20 ml 1X PBS at 4 °C for 5 min.
    (iii) 20 ml MQ water at 4 °C for 5 min.

18. After the final MQ water wash, blot filters upon a clean, dry Kimwipe. For each sample, divide filters into a small piece for microscopy and a large piece for cell sorting.

    ■ **Pause Point** Dry filters can be stored at −20 °C for several days.

Microscopy and Imaging •Timing 1 h

19. Drop 30 μl of DAPI working solution (1 μg ml^{-1}, as prepared in REAGENT SETUP) onto a Petri dish and place filter pieces onto droplets.
20. Incubate for 5 min in the dark.
21. Shortly wash first in MQ water, then 96% ethanol, before letting filters dry completely.
22. Place filter pieces on a microscope slide, microbe-side facing up. Add a sufficient volume of mounting medium (as prepared in REAGENT SETUP) and top with a coverslip.
23. Image the cells.

Cell Resuspension from Filter for Flow Cytometry •Timing 20 min

24. There are two effective methods for cell removal from FISH-ed filters (larger piece from step 18) for downstream cell sorting:

    (A) Vortexing

    (i) Cut the filter into pieces of approximately 1 cm^2 into a sterile Petri dish.
    (ii) Add filter pieces to 1 ml cell resuspension buffer (as prepared in REAGENT SETUP) in a 1.5-ml Eppendorf tube.
    (iii) Vortex at 4 °C for 15 min. Store supernatant at 4 °C for up to 3 days.

    (B) Sonication

    (i) Cut the filter into pieces of approximately 1 cm^2 into a sterile Petri dish.
    (ii) Add filter pieces to 1 ml 1X PBS in a 1.5-ml Eppendorf tube.
    (iii) Place the tube on ice and insert the sonication probe.
    (iv) Sonicate for 45 s (Bandelin Sonopuls HD70: 60% amplitude, pulser mode for 20 cycles).

Cell Sorting by Flow Cytometry •Timing 5 h
The wide variety of FACS, operational parameters, and environmental samples precludes a single detailed protocol for cell sorting. Below are general tips for sorting cells from environmental samples.

25. Install the tank with clean sheath fluid (as prepared in EQUIPMENT) and begin running.
26. Before sorting, UV-treat microtiter plates openly (without a cover) for 10 min. UV-treat plates again with a cover for another 10 min.
27. Filter each sample through an appropriate filter (e.g., 40-μm nylon mesh when using a 70-μm nozzle) to avoid clogging the FACS nozzle.
28. Stain cells with DAPI for 15 min in the dark. Generally, add 1–2 μl of DAPI stock solution (1 mg ml^{-1}) to 1 ml of resuspended cells.

    ▲**Critical Step** A wide variety of nucleic acid stains could be used at this step, and some work better than others for particular samples. The fluorescence labeling of the HCR–FISH hairpin probes should be considered and adapted if necessary.
29. Run the stained sample and define the desired microbial population with a sort gate (Fig. 4.4).

    ▲**Critical Step** It can be challenging to find the target population in some environmental samples. For example, samples with high cell density (>10^6 cells ml^{-1}) and low background signal show a distinct microbial population, whereas sediment samples generally display a less well-defined target. As

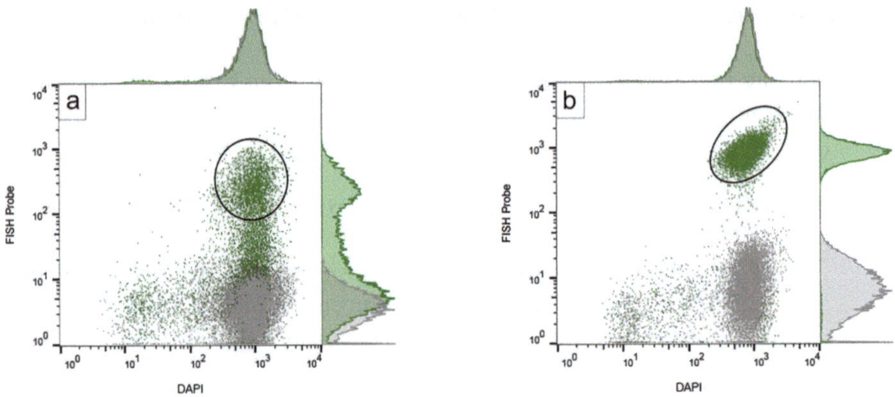

Fig. 4.4 The flow cytometry signals and sort windows (black oval) highlight a more clearly resolved labeled population in the fixed sample compared to the unfixed sample. (**a**) Unfixed *Escherichia coli* culture. (**b**) 70% ethanol-fixed *E. coli* culture. Gray, without initiator probe; green, with initiator probe

a general rule, run the controls to define the background (e.g., no initiator sample) before analyzing the hybridized samples.

30. Sort the target cells into the UV-treated microtiter plates.

    ▲**Critical Step** Accuracy of sorting is vital. We recommend diluting cells with sterile and filtered PBS ~10X before sorting and running the sample at a rate of approximately 50–500 cells/s, depending on the relative abundance of the target organism. In addition, the sort masks and logic of the instrument should be set to their most stringent feasible options.

31. **Negative Control**. Sort an equivalent volume of empty droplets into an empty well of the microtiter plate, e.g., 500 empty droplets if also sorting 500 cells per well. This sheath sample will undergo MDA amplification and sequencing if amplification products are generated to reveal potential contamination in the sheath fluid or FACS.

32. Cover plates with sterile foil and store at −80 °C.

    ■ **Pause Point** Sorted cells can be stored at −80 °C for up to several months.

4.2.5 Overall Timing

Day 1

Step 1, sample collection, preparation, and preservation: ~2 h.
Steps 2–9, probe hybridization: 3 h.
Steps 10–11, washing (1): 40 min.
Steps 12–16, hybridization chain reaction amplification: 2.5 h.
Steps 17–18, washing (2): 20 min.

Table 4.1 Troubleshooting table

Step	Problem	Possible reason	Solution
23	No or insufficient HCR–FISH signal		The use of helper oligonucleotides, which are unlabeled sequences flanking the targeting sequence, can increase the signal intensity (Fuchs et al. 2000).
		Insufficient cell permeabilization.	Gram positive cells can be treated for 5–10 min with a lysozyme solution (0.5–1 mg ml^{-1}), which facilitates the probe entry into the cell.
		Further optimization required.	If the probe is newly designed, it could be considered to lower the formamide concentration stepwise until a signal is detected.
23	Disrupted cells		The initial hybridization temperature can be reduced from 65 °C to 55 °C or skipped completely if necessary.
29	Cannot define the target population		A positive control by hybridization with a probe specific for a higher taxonomic level can facilitate narrowing down the signals of the target population in the flow cytometry scatter plot.

Day 2

Steps 19–23, microscopy and imaging: 1 h.
Step 24, cell resuspension from filter for flow cytometry: up to 20 min.
Steps 25–32, cell sorting by flow cytometry: 5 h.

4.2.6 Troubleshooting

Troubleshooting advice can be found in Table 4.1.

4.3 Anticipated Results

The described protocol enables the user to sort cells of a specific population based on known 16S rRNA gene sequences. Following DNA amplification, the resulting amplified products can be sequenced to recover genomes or specific genes. Throughout the protocol, the experimental success can be assessed at several stages.

4.3.1 Evaluation of a Successful Protocol

The initial assessment can be made after the HCR–FISH protocol part, where successfully hybridized cells can be visualized by fluorescence microscopy (step 23). This also provides information about whether the targeted population is attached to any particles or forming cell aggregates, which might hamper subsequent cell sorting. Additionally, this microscopic examination provides the relative abundance of the targeted population in the environmental sample if the proportion of target cells (with HCR–FISH label) to nontarget cells (DNA stain only) is measured.

The second assessment can be made during cell sorting by sorting the selected population (step 29) for microscopic analysis. This assessment will provide information about the purity of sorting by again counting target cells vs nontarget cells. A large proportion of cells should be successfully hybridized (>90%).

4.3.2 Further Analysis of the Results

Due to the low quantity of DNA in the low picogram range from 500 sorted cells, whole genome amplification is recommended, although direct Nextera library preparation may be applicable (Bowers et al. 2015; Rinke et al. 2016). The most often used whole genome amplification is multiple displacement amplification (MDA), due to its low amplification bias compared to other methods (Pinard et al. 2006). However, MDA is sensitive to DNA contamination, which can be mitigated through UV treatment of reagents, reaction plates, and pipette tips (Woyke et al. 2011), as well as reduction of the reaction volumes (Marcy et al. 2007), and increasing the amount of input DNA (Bergen et al. 2005).

Depending on the sorting purity, the MDA products may still contain nontarget sequences. The resulting sequence data thus requires careful analysis and quality assurance (Bowers et al. 2017). For whole genome analysis of the targeted population, we achieved good results applying metagenomic binning tools for separating target sequences from nontarget sequences after sequencing and assembly of the MDA products. This approach has yielded metagenome assembled genomes with completeness values up to 88% completeness and low contamination (<5%) (Grieb et al. 2020).

The identity of the sorted cells is determined by the specificity of the respective FISH probe, which has been used for flow cytometric sorting. In turn, the 16S rRNA gene sequence extracted from the genome assembly of the sorted cells should match with the probe sequence. For a more resolved taxonomic classification, phylogenetic trees can be calculated based on these 16S rRNA sequences, in addition to phylogenies based on concatenated alignments of protein-coding marker genes for subsequent comparison. Lastly, genomic assemblies will provide a foundation for metabolic predictions of the targeted taxa of interest.

Acknowledgments The work conducted by the US Department of Energy Joint Genome Institute (https://ror.org/04xm1d337), a DOE Office of Science User Facility, is supported by the Office of Science of the US Department of Energy, operated under Contract No. DE-AC02-05CH11231. This study was funded by the Max Planck Society.

Author Contributions This study was designed, directed, and coordinated by BF, RRM, and TW. AG, DG, JL, and MO carried out experiments. AG, DG, and JL drafted the manuscript with contributions from BF, RRM, and TW.

References

Albertsen M, Hugenholtz P, Skarshewski A, Nielsen KL, Tyson GW, Nielsen PH (2013) Genome sequences of rare, uncultured bacteria obtained by differential coverage binning of multiple metagenomes. Nat Biotechnol 31(6):533–538. https://doi.org/10.1038/nbt.2579

Amann R, Fuchs BM (2008) Single-cell identification in microbial communities by improved fluorescence in situ hybridization techniques. Nat Rev Microbiol 6(5):339–348. https://doi.org/10.1038/nrmicro1888

Anantharaman K, Brown CT, Hug LA, Sharon I, Castelle CJ, Probst AJ, Thomas BC, Singh A, Wilkins MJ, Karaoz U, Brodie EL, Williams KH, Hubbard SS, Banfield JF (2016) Thousands of microbial genomes shed light on interconnected biogeochemical processes in an aquifer system. Nat Commun 7(1):13219. https://doi.org/10.1038/ncomms13219

Barrero-Canosa J, Moraru C, Zeugner L, Fuchs BM, Amann R (2017) Direct-geneFISH: a simplified protocol for the simultaneous detection and quantification of genes and rRNA in microorganisms. Environ Microbiol 19(1):70–82. https://doi.org/10.1111/1462-2920.13432

Bergen AW, Qi Y, Haque KA, Welch RA, Chanock SJ (2005) Effects of DNA mass on multiple displacement whole genome amplification and genotyping performance. BMC Biotechnol 5(1):24. https://doi.org/10.1186/1472-6750-5-24

Berghuis BA, Yu FB, Schulz F, Blainey PC, Woyke T, Quake SR (2019) Hydrogenotrophic methanogenesis in archaeal phylum Verstraetearchaeota reveals the shared ancestry of all methanogens. Proc Natl Acad Sci 116(11):5037–5044. https://doi.org/10.1073/pnas.1815631116

Bowers RM, Clum A, Tice H, Lim J, Singh K, Ciobanu D, Ngan CY, Cheng J-F, Tringe SG, Woyke T (2015) Impact of library preparation protocols and template quantity on the metagenomic reconstruction of a mock microbial community. BMC Genomics 16:856. https://doi.org/10.1186/s12864-015-2063-6

Bowers RM, Doud DFR, Woyke T (2017) Analysis of single-cell genome sequences of bacteria and archaea. Emerg Top Life Sci 1(3):125–131. https://doi.org/10.1042/ETLS20160028

Bowers RM, Nayfach S, Schulz F, Jungbluth SP, Ruhl IA, Sheremet A, Lee J, Goudeau D, Eloe-Fadrosh EA, Stepanauskas R, Malmstrom RR, Kyrpides NC, Dunfield PF, Woyke T (2022) Dissecting the dominant hot spring microbial populations based on community-wide sampling at single-cell genomic resolution. ISME J 16(5):1337–1347. https://doi.org/10.1038/s41396-021-01178-4

Bruder LM, Dörkes M, Fuchs BM, Ludwig W, Liebl W (2016) Flow cytometric sorting of fecal bacteria after in situ hybridization with polynucleotide probes. Syst Appl Microbiol 39(7):464–475. https://doi.org/10.1016/j.syapm.2016.08.005

Campbell AG, Campbell JH, Schwientek P, Woyke T, Sczyrba A, Allman S, Beall CJ, Griffen A, Leys E, Podar M (2013) Multiple single-cell genomes provide insight into functions of uncultured Deltaproteobacteria in the human oral cavity. PLoS One 8(3):e59361. https://doi.org/10.1371/journal.pone.0059361

Chen L-X, Anantharaman K, Shaiber A, Eren AM, Banfield JF (2020) Accurate and complete genomes from metagenomes. Genome Res 30(3):315–333. https://doi.org/10.1101/gr.258640.119

Clingenpeel S, Schwientek P, Hugenholtz P, Woyke T (2014) Effects of sample treatments on genome recovery via single-cell genomics. ISME J 8(12):2546–2549. https://doi.org/10.1038/ismej.2014.92

Dam HT, Vollmers J, Sobol MS, Cabezas A, Kaster A-K (2020) Targeted cell sorting combined with single cell genomics captures low abundant microbial dark matter with higher sensitivity than metagenomics. Front Microbiol 11:1377. https://doi.org/10.3389/fmicb.2020.01377

Doud DFR, Bowers RM, Schulz F, De Raad M, Deng K, Tarver A, Glasgow E, Vander Meulen K, Fox B, Deutsch S, Yoshikuni Y, Northen T, Hedlund BP, Singer SW, Ivanova N, Woyke T (2020) Function-driven single-cell genomics uncovers cellulose-degrading bacteria from the rare biosphere. ISME J 14(3):659–675. https://doi.org/10.1038/s41396-019-0557-y

Engel P, Stepanauskas R, Moran NA (2014) Hidden diversity in honey bee gut symbionts detected by single-cell genomics. PLoS Genet 10(9):e1004596. https://doi.org/10.1371/journal.pgen.1004596

Fuchs BM, Glöckner FO, Wulf J, Amann R (2000) Unlabeled helper oligonucleotides increase the in situ accessibility to 16S rRNA of fluorescently labeled oligonucleotide probes. Appl Environ Microbiol 66(8):3603–3607. https://doi.org/10.1128/AEM.66.8.3603-3607.2000

Glöckner FO, Yilmaz P, Quast C, Gerken J, Beccati A, Ciuprina A, Bruns G, Yarza P, Peplies J, Westram R, Ludwig W (2017) 25 years of serving the community with ribosomal RNA gene reference databases and tools. J Biotechnol 261(Supplement C):169–176. https://doi.org/10.1016/j.jbiotec.2017.06.1198

Greuter D, Loy A, Horn M, Rattei T (2016) probeBase—an online resource for rRNA-targeted oligonucleotide probes and primers: new features 2016. Nucleic Acids Res 44(D1):D586–D589. https://doi.org/10.1093/nar/gkv1232

Grieb A, Bowers RM, Oggerin M, Goudeau D, Lee J, Malmstrom RR, Woyke T, Fuchs BM (2020) A pipeline for targeted metagenomics of environmental bacteria. Microbiome 8(1):21. https://doi.org/10.1186/s40168-020-0790-7

Hausmann B, Knorr K-H, Schreck K, Tringe SG, Glavina del Rio T, Loy A, Pester M (2016) Consortia of low-abundance bacteria drive sulfate reduction-dependent degradation of fermentation products in peat soil microcosms. ISME J 10(10):2365–2375. https://doi.org/10.1038/ismej.2016.42

Ji P, Zhang Y, Wang J, Zhao F (2017) MetaSort untangles metagenome assembly by reducing microbial community complexity. Nat Commun 8:14306. https://doi.org/10.1038/ncomms14306

Jing X, Gou H, Gong Y, Su X, Xu L, Ji Y, Song Y, Thompson IP, Xu J, Huang WE (2018) Raman-activated cell sorting and metagenomic sequencing revealing carbon-fixing bacteria in the ocean. Environ Microbiol 20(6):2241–2255. https://doi.org/10.1111/1462-2920.14268

Jousset A, Bienhold C, Chatzinotas A, Gallien L, Gobet A, Kurm V, Küsel K, Rillig MC, Rivett DW, Salles JF, van der Heijden MGA, Youssef NH, Zhang X, Wei Z, Hol WHG (2017) Where less may be more: how the rare biosphere pulls ecosystems strings. ISME J 11(4):853–862. https://doi.org/10.1038/ismej.2016.174

Kalyuzhnaya MG, Zabinsky R, Bowerman S, Baker DR, Lidstrom ME, Chistoserdova L (2006) Fluorescence in situ hybridization-flow cytometry-cell sorting-based method for separation and enrichment of type I and type II methanotroph populations. Appl Environ Microbiol 72(6):4293–4301. https://doi.org/10.1128/AEM.00161-06

Kashtan N, Roggensack SE, Rodrigue S, Thompson JW, Biller SJ, Coe A, Ding H, Marttinen P, Malmstrom RR, Stocker R, Follows MJ, Stepanauskas R, Chisholm SW (2014) Single-cell genomics reveals hundreds of coexisting subpopulations in wild Prochlorococcus. Science 344(6182):416–420. https://doi.org/10.1126/science.1248575

Ludwig W, Strunk O, Westram R, Richter L, Meier H, Yadhukumar, Buchner A, Lai T, Steppi S, Jobb G, Forster W, Brettske I, Gerber S, Ginhart AW, Gross O, Grumann S, Hermann S, Jost R, Konig A, Liss T, Lussmann R, May M, Nonhoff B, Reichel B, Strehlow R, Stamatakis A, Stuckmann N, Vilbig A, Lenke M, Ludwig T, Bode A, Schleifer K-H (2004) ARB: a software environment for sequence data. Nucleic Acids Res 32(4):1363–1371. https://doi.org/10.1093/nar/gkh293

Lynch MDJ, Neufeld JD (2015) Ecology and exploration of the rare biosphere. Nat Rev Microbiol 13(4):217–229. https://doi.org/10.1038/nrmicro3400

Manz W, Amann R, Ludwig W, Wagner M, Schleifer K-H (1992) Phylogenetic oligodeoxynucleotide probes for the major subclasses of proteobacteria: problems and solutions. Syst Appl Microbiol 15:593–600. https://doi.org/10.1016/S0723-2020(11)80121-9

Marcy Y, Ishoey T, Lasken RS, Stockwell TB, Walenz BP, Halpern AL, Beeson KY, Goldberg SMD, Quake SR (2007) Nanoliter reactors improve multiple displacement amplification of genomes from single cells. PLoS Genet 3(9):e155. https://doi.org/10.1371/journal.pgen.0030155

Miyauchi R, Oki K, Aoi Y, Tsuneda S (2007) Diversity of nitrite reductase genes in Candidatus Accumulibacter phosphatis-dominated cultures enriched by flow-cytometric sorting. Appl Environ Microbiol 73(16):5331–5337. https://doi.org/10.1128/AEM.00175-07

Pernthaler A, Pernthaler J, Amann R (2002) Fluorescence in situ hybridization and catalyzed reporter deposition for the identification of marine bacteria. Appl Environ Microbiol 68(6):3094–3101. https://doi.org/10.1128/AEM.68.6.3094-3101.2002

Pinard R, de Winter A, Sarkis GJ, Gerstein MB, Tartaro KR, Plant RN, Egholm M, Rothberg JM, Leamon JH (2006) Assessment of whole genome amplification-induced bias through high-throughput, massively parallel whole genome sequencing. BMC Genomics 7(1):216. https://doi.org/10.1186/1471-2164-7-216

Podar M, Abulencia CB, Walcher M, Hutchison D, Zengler K, Garcia JA, Holland T, Cotton D, Hauser L, Keller M (2007) Targeted access to the genomes of low-abundance organisms in complex microbial communities. Appl Environ Microbiol 73(10):3205–3214. https://doi.org/10.1128/AEM.02985-06

Pratscher J, Vollmers J, Wiegand S, Dumont MG, Kaster A-K (2018) Unravelling the identity, metabolic potential and global biogeography of the atmospheric methane-oxidizing upland soil cluster α. Environ Microbiol 20(3):1016–1029. https://doi.org/10.1111/1462-2920.14036

Quast C, Pruesse E, Yilmaz P, Gerken J, Schweer T, Yarza P, Peplies J, Glöckner FO (2013) The SILVA ribosomal RNA gene database project: improved data processing and web-based tools. Nucleic Acids Res 41(D1):D590–D596. https://doi.org/10.1093/nar/gks1219

Rinke C (2018) Single-cell genomics of microbial dark matter. Methods Mol Biol 1849:99–111. https://doi.org/10.1007/978-1-4939-8728-3_7

Rinke C, Schwientek P, Sczyrba A, Ivanova NN, Anderson IJ, Cheng J-F, Darling A, Malfatti S, Swan BK, Gies EA, Dodsworth JA, Hedlund BP, Tsiamis G, Sievert SM, Liu W-T, Eisen JA, Hallam SJ, Kyrpides NC, Stepanauskas R, Rubin EM, Hugenholtz P, Woyke T (2013) Insights into the phylogeny and coding potential of microbial dark matter. Nature 499:431–437. https://doi.org/10.1038/nature12352

Rinke C, Lee J, Nath N, Goudeau D, Thompson B, Poulton N, Dmitrieff E, Malmstrom R, Stepanauskas R, Woyke T (2014) Obtaining genomes from uncultivated environmental microorganisms using FACS–based single-cell genomics. Nat Protoc 9(5):1038–1048. https://doi.org/10.1038/nprot.2014.067

Rinke C, Low S, Woodcroft BJ, Raina J-B, Skarshewski A, Le XH, Butler MK, Stocker R, Seymour J, Tyson GW, Hugenholtz P (2016) Validation of picogram- and femtogram-input DNA libraries for microscale metagenomics. PeerJ 4:e2486. https://doi.org/10.7717/peerj.2486

Schulz F, Alteio L, Goudeau D, Ryan EM, Yu FB, Malmstrom RR, Blanchard J, Woyke T (2018) Hidden diversity of soil giant viruses. Nat Commun 9(1):4881. https://doi.org/10.1038/s41467-018-07335-2

Sekar R, Fuchs BM, Amann R, Pernthaler J (2004) Flow sorting of marine bacterioplankton after fluorescence in situ hybridization. Appl Environ Microbiol 70(10):6210–6219. https://doi.org/10.1128/AEM.70.10.6210-6219.2004

Wallner G, Amann R, Beisker W (1993) Optimizing fluorescent in situ hybridization with rRNA-targeted oligonucleotide probes for flow cytometric identification of microorganisms. Cytometry 14:136–143. https://doi.org/10.1002/cyto.990140205

Wallner G, Fuchs B, Spring S, Beisker W, Amann R (1997) Flow sorting of microorganisms for molecular analysis. Appl Environ Microbiol 63(11):4223–4231. https://doi.org/10.1128/aem.63.11.4223-4231.1997

Wang Y, Hatt JK, Tsementzi D, Rodriguez-R LM, Ruiz-Pérez CA, Weigand MR, Kizer H, Maresca G, Krishnan R, Poretsky R, Spain JC, Konstantinidis KT (2017) Quantifying the importance of the rare biosphere for microbial community response to organic pollutants in a freshwater ecosystem. Appl Environ Microbiol 83(8):e03321-16. https://doi.org/10.1128/AEM.03321-16

Woyke T, Sczyrba A, Lee J, Rinke C, Tighe D, Clingenpeel S, Malmstrom R, Stepanauskas R, Cheng J-F (2011) Decontamination of MDA reagents for single cell whole genome amplification. PLoS One 6(10):e26161. https://doi.org/10.1371/journal.pone.0026161

Woyke T, Doud DFR, Schulz F (2017) The trajectory of microbial single-cell sequencing. Nat Methods 14(11):1045. https://doi.org/10.1038/nmeth.4469

Yamaguchi T, Fuchs BM, Amann R, Kawakami S, Kubota K, Hatamoto M, Yamaguchi T (2015) Rapid and sensitive identification of marine bacteria by an improved in situ DNA hybridization chain reaction (quickHCR-FISH). Syst Appl Microbiol 38(6):400–405. https://doi.org/10.1016/j.syapm.2015.06.007

Yarza P, Yilmaz P, Pruesse E, Glöckner FO, Ludwig W, Schleifer K-H, Whitman WB, Euzéby J, Amann R, Rosselló-Móra R (2014) Uniting the classification of cultured and uncultured bacteria and archaea using 16S rRNA gene sequences. Nat Rev Microbiol 12(9):635–645. https://doi.org/10.1038/nrmicro3330

Yilmaz S, Haroon MF, Rabkin BA, Tyson GW, Hugenholtz P (2010) Fixation-free fluorescence in situ hybridization for targeted enrichment of microbial populations. ISME J 4(10):1352–1356. https://doi.org/10.1038/ismej.2010.73

Yu FB, Blainey PC, Schulz F, Woyke T, Horowitz MA, Quake SR (2017) Microfluidic-based mini-metagenomics enables discovery of novel microbial lineages from complex environmental samples. elife 6:e26580. https://doi.org/10.7554/eLife.26580

5 Advancements in the Application of Activity-Based Studies for Separation of Microbial Populations via Fluorescence-Activated Cell Sorting

Nicholas J. Reichart

Abstract

Activity-based studies allow for the separation of specific phenotypes expressed by microbes from complex environments. This chapter discusses the application of two bioorthogonal approaches for the labeling and enrichment of microbes across a variety of ecosystems. Sample processing steps and concerns are described to assist in the successful implementation of these studies.

Keywords

Genomics · Fluorescence-activated cell sorting · Microbial Physiology · Phenotype · Bioorthogonal

5.1 Introduction

Microbial phenotypes cannot be determined solely by genomic sequencing as the phenome is dependent on environmental conditions and biotic interactions (Jansson and Hofmockel 2018) and requires direct activity measurements. Deciphering a clear understanding of microbial activities allows for the development of cultivation conditions for uncultured microbes or refinements in models of global nutrient cycling. In this chapter, we discuss the growing discipline of activity-based studies that are suited to label and enrich microbial populations with specific functions or phenotypes to enhance our knowledge of both cultured and uncultured microbes.

N. J. Reichart (✉)
Earth and Biological Sciences Directorate, Pacific Northwest National Laboratory, Richland, WA, USA
e-mail: nicholas.reichart@pnnl.gov

C. Rinke, M. Džunková (eds.), *Single-Cell Omics in Microbiome Research*, The Microbiomes of Humans, Animals, Plants, and the Environment 7,
https://doi.org/10.1007/978-3-032-07527-7_5

Additionally, examples of the application of activity studies across various ecosystems will be further explored.

Microbes exist in complex arrangements and structures, whether in environmental niches within soil or sediment matrices, in bioreactors with regions of heterogeneity, or throughout the human microbiome. Genomic advancements throughout the previous decades have assisted researchers in being able to detect and determine the DNA composition of microbes, increasing the breadth of known microbes in databases (Hug 2024; Hug et al. 2016; Rinke et al. 2013). Amplicon studies of the microbial 16S rRNA gene and shotgun metagenomic surveys have uncovered myriad novel taxa. However, genomic screening methods cannot differentiate between living, functionally active biomass from extracellular DNA or relic, dead microbes, which can comprise, on average, 40% of the microbial richness across various soils (Carini et al. 2016). As the number of known taxa continues to increase, the connection to microbial function lags. This is further confounded by heterogeneous populations within individual lineages or across micro and macro spatial scales. Microbial activity is difficult to determine for environmental microbes and may not always directly relate to the genomic potential uncovered via DNA sequencing. In situ microbial activity is obscured behind vast metabolic networks or difficult-to-parse matrices, and thus, these bulk sequencing methods miss the connection between taxonomic identification with expressed phenotype.

Extracting microbes from environmental matrices and obtaining physiology measurements to build a connection from functional potential to in situ function presents unique challenges for each ecosystem type. Determining the molecular basis driving the phenotype has emerged as a powerful approach to work in reverse to assign the phenotype to the genomic framework. This involves identifying a phenotype or activity in microbial populations and then establishing which microbes are present and what machinery or genomic potential these microbial populations express to develop the molecular basis of biological function. Genomic hypotheses are essential for inferring potential, but linking the in situ function to the genomic repertoire or expressed genes provides a definitive view of microbial function. Specific interrogation of the qualities driving phenotypes of interest can be ascertained following the separation of a microbial population based on a functional profile. By examining two widely used activity-focused approaches, bioorthogonal non-canonical amino acid tagging and activity-based protein profiling, microbial phenomics through activity studies will be explored (Fig. 5.1). These methods, part of substrate analogue probing (Hatzenpichler et al. 2020), are two of a wide suite of activity approaches to bridge the disconnect of genome to phenome that focuses on the expression of function and serves as the topic of this chapter.

5.2 Enrichment Through Flow Cytometry

While flow cytometry has been used to quantify microbial populations, fluorescent-activated cell sorting (FACS) has the unique ability to separate populations based on levels of activity through the incorporation of substrate analogue probes and, in the

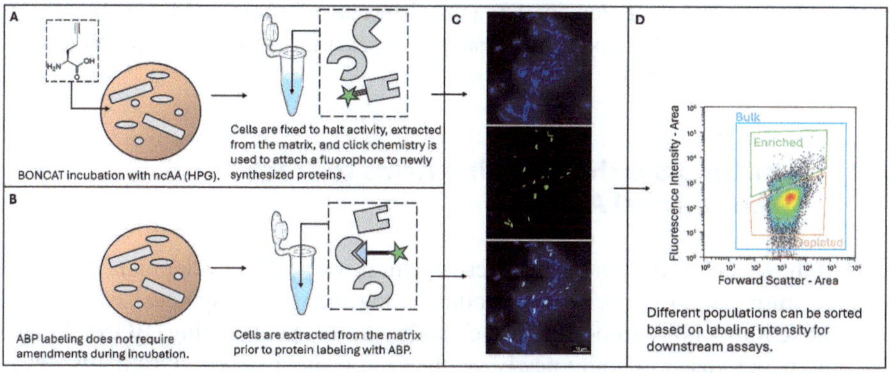

Fig. 5.1 Comparison of BONCAT and ABP methods. (**a**) BONCAT-based experiments utilize the incorporation of a noncanonical amino acid (HPG) into newly synthesized proteins. Protein synthesis is stopped through the fixation of samples to achieve a snapshot of activity within the incubations. Cells are extracted from the sample matrix, and newly synthesized proteins are labeled with a fluorophore via click chemistry. (**b**) ABP processing does not require any compounds to be added during the sample incubation. Once cells are extracted from the sample matrix, the ABP is incubated with the cells and interacts with protein targets of interest. (**c**) Microscopy can be used to visualize BONCAT or ABP incorporation into cells. DAPI (cyan) is used as a counterstain to identify total microbial populations. Activity labeling (green) occurs in a subset of cells and with varying levels of intensity depending on the activity of the sample. An overlay is used to compare labeled cell numbers to total cell population counts. Images taken with 63x objective; scale bar represents 10 μm. (**d**) FACS processing to separate cell populations based on labeling intensity (bulk = total cell population, depleted = unlabeled cells, enriched = labeled cells) allows for downstream assays, including activity incubation or genomic sequencing. Once labeled, samples from BONCAT or ABP incubation follow a similar processing workflow

cases described within this chapter, coupled with click chemistry to attach a fluorescent dye. For FACS processing, a fluorogenic dye is typically attached as the marker, enabling separation of labeled microbial populations based on the fluorescent signal of the bound dye. There are a variety of potential fluorophores that can be adjusted to fit the necessary fluorogenic panels for each study. Counterstain labeling can be used to label nucleic acids to be able to select for biological material (total or viable cells) before sorting based on the activity marker of interest. Care needs to be taken to optimize the fluorophore to the instruments being used as microscopes and cytometers can vary in terms of excitation lasers and filters for detection.

When applying FACS to environmental samples, such as soil, various wash methods have been employed to extract the cells from the native matrix (Khalili et al. 2019). A clean sample is required to prevent clogging issues with FACS and to provide adequate separation of labeled events to be able to clearly distinguish cell events from debris events. Brief vortexing with a mild detergent (e.g., 0.02% Tween20) is often used to detach cells from environmental matrices, followed by slow-speed centrifugation to pellet debris or filtration, leaving cells in solution (Couradeau et al. 2019; Reichart et al. 2020). Other methods to detach cells from environmental particles include using density gradients (Krukenberg et al. 2021; Reichart et al. 2023) or sonication (Du and Behrens 2021). Due to potential biases in sample extraction efficiencies towards different microbial clades, subsampling before and after wash steps is necessary to track the exclusion of potential taxa (Krukenberg et al. 2021).

5.3 Utilizing Synthetic Amino Acids to Track Translational Activity

One of the recent advancements for activity studies was the development of the use of non-canonical amino acids to introduce a tag into newly synthesized proteins. The application of bioorthogonal non-canonical amino acid tagging (BONCAT) has become widely used to study newly synthesized protein targets. Originally developed in mammalian cells to explore rapid responses of cells to environmental cues, BONCAT relies on natural tRNA synthetases in cells that incorporate the non-canonical amino acid (ncAA) that is analogous to methionine (Dieterich et al. 2006). The strength of BONCAT relies on its ability to distinguish newly synthesized proteins from the preexisting protein pool and can be used as a proxy for phenotype expression.

The terminal alkyne or azide group present on the methionine analogs, L-homopropargylglycine (HPG) or L-azidohomoalanine (AHA), respectively, allows researchers to perform click chemistry (Kiick et al. 2001) to attach a marker of interest to the newly synthesized protein. In brief, the terminal group on the ncAA reacts with the group on the tagging molecule in the presence of a ligand to generate a triazole ring linking the two groups together in a covalent bond. Early studies with BONCAT focused on protein enrichment through click chemistry of the proteins to detect changes in the proteome via gel imaging or through biotin-streptavidin-conjugated enrichment for mass spectrometry-based proteomics. Expanding beyond proteome labeling, BONCAT is amendable to enrichment of the active whole-cell microbial populations through coupling with flow cytometry and fluorescence-activated cell sorting (FACS).

BONCAT has been tested in a variety of systems and the effects on sample toxicity have been screened. Originating from the work of Dieterich and colleagues in 2006, BONCAT toxicity was tested through screening with the synthetic amino acid and methionine as the control substrate. No differences in cell viability or protein synthesis quantity were detected within the 2-hour incubation window tested (Dieterich et al. 2006). However, for longer incubations, impacts on cell growth of *E. coli* could be detected with both AHA or HPG at different times and concentrations (Landor et al. 2023). For applications with phytoplankton, growth of *Synochoccocus* was inhibited at higher concentrations of HPG. At lower concentrations of HPG that did not alter growth rates, incubations with nitrogen-limited stress magnified the proteomic changes compared to control incubations (Michels et al. 2021). Additionally, a study on the metabolomic response of HPG incubation found no detrimental effect on *E. coli* under stable incubation conditions; however, when heat-stressed, the global metabolome was affected within a generation time (Steward et al. 2020). While these impacts can be directly observed on axenic lab-based cultures, both the incorporation of the ncAA and the effects on growth become difficult to benchmark for environmental samples when the total population and composition are unknown. For other sample types tested, typically, researchers utilize unamended samples compared with analog incubated samples to screen for impacts of the synthetic amino acid. Frequent sampling for sequencing aids in the evaluation of the detriment or lack thereof to the ecosystem's microbial populations. Incubations for BONCAT are inherently a destructive process, requiring fixation to freeze translational activity at a specific point in time after sampling. Long persistent incubations could lead to crossfeeding and transfer of synthetic amino acids or cause adverse growth conditions as most cultured experiments with BONCAT are completed within one generation time. As such, when multiple time windows are of interest for an experiment, separate incubations must be set up to minimize ncAA exposure and to capture specific time points.

The separation of microbes of interest from the total cell population allows for various downstream assays. Activity screening via BONCAT-FACS is a high-throughput method, ultimately relying on adequate quantities of cell numbers and fluorescent labeling, allowing for a large number of conditions to be tested, along with sorting upwards of 100,000 events per sample. Through taxonomic identification following BONCAT-FACS, microbial populations can be compared for differential abundance analyses to determine regulation of activity when compared to a baseline sample activity. As BONCAT is nondiscriminatory in labeling, the timing of the incubation must be adjusted to achieve sufficient labeling without saturating the system. Typically, BONCAT activity studies coupled with FACS aim to describe the microbial populations using either 16S rRNA gene sequencing (Couradeau et al. 2019; Reichart et al. 2020; Krukenberg et al. 2021) or metagenomic sequencing (McKay et al. 2022; Schweitzer et al. 2022; Trexler et al. 2023). In the following section, we highlight recent BONCAT activity studies that have progressed toward linking microbial activity with taxonomic identifications.

5.4 BONCAT Development into Microbial Systems

BONCAT is inherently a protein screening method that was expanded to detect metabolically active intact cells in a high-throughput fashion. Early applications of BONCAT for microbial populations utilized the incorporation of the synthetic amino acid L-Azidohomoalanine (AHA) to label pure cultures and multiple different environmental samples, including the oral microbiome, sediment, and freshwater (Hatzenpichler et al. 2014). Fluorescence in situ hybridization (FISH) was used to connect activity to taxonomy, linking the translational activity of BONCAT to microbial identification. Labeling of *E. coli* with BONCAT was detected in as short as a two-minute incubation, highlighting the sensitivity of the method. Due to the rapid labeling with BONCAT, fixation through crosslinking proteins via formaldehyde is generally applied to arrest continued protein synthesis. Further testing of the initial application of BONCAT for microbial cells showed strong signal incorporation for methanotrophic cultures only in the presence of methane and limited labeling in methane-replete samples (Hatzenpichler et al. 2014).

Hatzenpichler and colleagues continued to innovate with the application of BONCAT towards microbial populations, later developing a workflow to apply BONCAT in conjunction with FACS to enrich populations of translationally active microbes (Hatzenpichler et al. 2016). Through the incorporation of FACS into the workflow, translationally active cells were separated from complex samples. In this study, anaerobic methane-oxidizing archaea (ANME) and sulfate-reducing bacteria (SRB) consortia were BONCAT labeled, sorted, and sequenced for metagenomics and 16S rRNA gene identification. The ANME-SRB consortia are slow-growing with multi-month generation times, and despite this, the BONCAT signal was able to be detected in as short as 5 days of incubation. BONCAT incubations are nondiscriminatory and label any translational activity that is occurring. Reagent specificity, as is used for stable isotope probing (SIP), is not needed and allows for BONCAT interrogation into unique systems like ANME/AOB consortia regardless of substrate amendment or metabolic processes.

Moving into other ecosystems, the application of BONCAT-FACS for genomic sequencing was used to describe the active fraction of the soil microbiome. Soil processes have a strong impact on global element cycling and models for carbon turnover and greenhouse gas production. Determining levels of microbial activity in soil ecosystems aids in more accurate modeling of soil functions and health (Meurer et al. 2020; Vereecken et al. 2016). Couradeau and colleagues applied BONCAT–FACS to two depth profiles of a soil system to determine the active fraction and the microbial populations responsible for general translational activities (Couradeau et al. 2019). The active fraction detected via BONCAT–FACS decreased across depth profiles suggesting models of soil function must take depth of soil horizons into account. Additionally, the active population exists as a unique fraction of the soil microbiome when compared to the total microbial community, which was typically queried using total or bulk DNA profiling. By applying BONCAT-FACS as an enrichment protocol for soil microbes, the ecological information retrieved is relevant for active members within the communities and separate from dormant or dead microbes that would be detected without first enrichment.

Spatial arrangement of active microbes can be tracked across a depth gradient as well as in association with soil minerals using BONCAT. In a benchmark study to couple the spatial configuration of active organisms with mineralogy, a coupled approach of BONCAT–FACS and correlative fluorescence microscopy in resin-embedded cores was used to describe microbial mineral relationships with levels of activity and taxonomic identification (Marlow et al. 2021). BONCAT labeling and activity decreased with increasing depth profiles as oxygen became depleted. Additionally, by correlative microscopy, microbial activity in relation to minerals showed higher amounts of active cells associated with non-quartz minerals.

BONCAT–FACS is a method readily adaptable to substrate screening. The low initial costs of reagents and the broad applicability of the synthetic amino acid incorporation increase the general usage of this method over more specified techniques, such as SIP. Multiple publications have emerged to screen environmental samples for substrate utilization by the native microbial communities. Reichart et al. 2020 applied BONCAT-based incubations to screen for metabolic activity of hot spring sediment slurries for 25 substrate amendments or conditional changes. BONCAT incubations are amendable to a range of substrate types (e.g., leucine, cellobiose, pyruvate, etc.), concentration gradients, and headspace compositions (microoxic, anoxic) (Reichart et al. 2020). While this study saw minimal changes among many of the conditions tested compared to the baseline measurement, incubations containing cellobiose or an anoxic headspace had significant shifts in translationally active microbial populations. Combining a range of incubation lengths in the presence of HPG could help to distinguish fast-growing microbes or taxa that rely on metabolic crossfeeding within a system.

Additional substrate screening utilized deep-sea sediments from the Guaymas Basin to analyze the degradation of a variety of polysaccharides (Krukenberg et al. 2021). These compounds accumulate on the sea floor and can be major spots for carbon turnover in the hydrothermal vent ecosystems. Members within *Thermotogae* were detected as abundant populations in the active cell sorted fraction across all substrates screened, whereas *Clostridia* were mostly enriched in samples with laminarin or starch. BONCAT works as a screening tool for cultured and uncultured microbes throughout various environmental systems. This method has been shown to be effective as a high-throughput method for detecting microbial responses to changing conditions and substrate amendment.

Decomposition of complex compounds occurs in phases with an interplay of division of labor and crossfeeding between key degrading or scavenger organisms (Reintjes et al. 2019; Wilhelm et al. 2019). To gain an understanding of early responders and later responders, HPG can be added at different time points in separate incubations. This allows for snapshots of activity across discrete phases of an incubation. In one such incubation investigating the decomposition of plant biomass, microbial diversity initially decreases with carbon input but increases later during the incubations, suggesting an increase of metabolic strategies available to a broader microbial population in the late phase of decomposition (Frates et al. 2023). Tracking the relative abundance of active microbial populations over time offers evidence for microbial metabolism patterns and effects of changing conditions in environmental samples.

The bioeconomy is another strong area for the use of BONCAT–FACS, in particular, activated sludge wastewater systems. The function of wastewater reactors relies heavily on microbial activity to consume, break down, and cycle elements. Improvements to the processing or engineered alterations in response to changing conditions would provide beneficial advancements to the overall efficiency of treatment plants. Following a time course of a press disturbance for additions of free ammonia, the activity of nitrite-oxidizing bacterial populations was tracked. The differences in microbial community composition peaked at 75 days postinoculation and following the bioreactor incubations, microcosms with HPG were inoculated to detect differentially active microbes (Madill et al. 2021). The portion of translationally active microbes was lower than expected steady-state modeling, suggesting a decrease in performance of the bioreactors. The microbes that were active helped to resolve the taxa performing the functions in the bioreactors following long-term incubations after a free ammonia disturbance.

Human health applications are also relevant for the application of BONCAT-FACS to be put into practice. Disease state may vary in levels of activity related to the expression of symptoms within patients. Being able to track relevant microbiota can help to identify causes of disease or deficiencies. Cystic fibrosis (CF) lung infection is a chronic disease extenuating by colonization of pathogenic bacteria including *Pseudomonas*. Valentini and colleagues applied BONCAT-FACS to decipher the percentage of active microbes in the sputum of CF patients (Valentini et al. 2020). While the percentage of active microbes detected varied heavily for different patients, a *Pseudomonas* was consistently detected as a higher abundance in CF-expressing patients. Determining active microbes within a CF infection could lead to personalized medical intervention to treat against specific microbes within a patient. Interestingly, similar taxa were detected in both the BONCAT positive and negative FACS fractions suggesting heterologous populations of taxa within a sample. This occurrence of mixed levels of activity has been noticed across multiple BONCAT studies (Couradeau et al. 2019; Trexler et al. 2023) and requires further methods and interrogation to fully decipher the functional markers which perpetuate populations of cell heterogeneity.

Community profiling for functionally active microbes has been approached in several ways to compare against the total microbiome. In a study by Taguer et al., microbial communities retrieved from fecal samples were profiled to compare the composition of active members for nucleic acid content and protein translation (Taguer et al. 2021). High nucleic acid populations, as determined by counterstaining of all nucleic acid content, were a distinct population compared to the BONCAT, translationally active populations which were a subset of the total community. As a result, BONCAT represents a valid alternative to determining bacterial activity with a direct readout of translational function without the need for DNA replication.

Building on the genomic information gathered from BONCAT–FACS-Seq using amplicon sequencing of the 16S rRNA gene, several studies have begun utilizing BONCAT–FACS to sequence for metagenomics (McKay et al. 2022; Schweitzer et al. 2022; Trexler et al. 2023). McKay and Schweitzer (McKay et al. 2022; Schweitzer et al. 2022) applied BONCAT–FACS mini-metagenomes to describe the functional potential of active microbes from deep subsurface microbial populations with limited available biomass. In this study, amplification of cell aliquots of 300 and 5000 sorted cells was used to generate mini-metagenomes. Using BONCAT-FACS to first enrich for active microbes, two metagenome assembled genomes (MAGs) were recovered that were not broadly detected via shotgun metagenomics. This provided a functional enrichment phase prior to sequencing that allowed for a more detailed interrogation into the phenotypes within coalbed methane microbial communities.

In later applications of BONCAT–FACS for metagenomes, Trexler and colleagues incubated biocrust samples following rewetting events for two distinct time frames to detect early and late window microbial responses (Trexler et al. 2023). By analyzing microbial responses in activity over two incubation lengths, early and late responders to the rewetting event could be described. As seen with previous soil studies, the active fraction of the biocrust samples is a subset of the total population and is unique from the inactive fraction that lacks BONCAT labeling. At the different times studied here, microbial populations of *Chitinophagaceae* and families within Bacillota were more active following rewetting, suggesting important contributors to the response function within the biocrust. Combining BONCAT–FACS with metagenomes also allowed detection of enriched genomic potential, which could be contributing to the expressed phenotype within active populations. At both the short- and long-term incubation windows, genes related to cell motility were detected as enriched in the active fraction. BONCAT–FACS generated MAGs directly connect the active microbiome to a potential function existing in phenotype-enriched populations.

Here, we have highlighted a brief recap of BONCAT studies and experiments that have linked the metabolic labeling of translationally active microbes to taxonomy through microbial DNA sequencing via FACS enrichment. Continual developments applying BONCAT are pushing this approach to achieve more specificity and utility in the environmental, industrial, and health-associated fields.

5.5 Activity-Based Probes for Microbial Identification

In this next section, we will be describing work associated with activity-based protein profiling, a method to label specific, active proteins linked to a phenotype or expression profile of interest. Activity-based protein profiling (ABPP) employs activity-based probes (ABPs), small molecular structures designed to mimic enzymatic substrates. ABPs have been used to elucidate biochemical pathways and annotate proteins of unknown functions (Barglow and Cravatt 2007) and can be designed to a variety of protein targets ranging from organic matter (Zegeye et al. 2021; Chauvigne-Hines et al. 2012; Schroder et al. 2019), virulence factors (Krishnamoorthy et al. 2022), or metabolite products (Webber et al. 2023). An activity-based probe is composed of three regions: a reactive chemical moiety, a binding group, and a marker tag region. The reactive chemical moiety creates a covalent bond linking the probe to the target. The binding group is the structure uniquely designed to mimic the metabolite or compound that interacts with the protein of interest. Third, the marker tag region is an alkyne or azide group that can be used to attach reporters via click chemistry. Similar to BONCAT, these reporters can be fluorophores for visualization or biotin for capture-based enrichment. While the components of ABPs are important for the understanding of their application, the process of designing and synthesizing ABPs is beyond the scope of this chapter and has been reviewed extensively elsewhere (Barglow and Cravatt 2007; Cravatt et al. 2008; Fang et al. 2021; Willems et al. 2011). Nonetheless, the specificity gained in a well-designed ABP increases the effectiveness towards identifying microbial phenotypes within mixed community samples. Much like the origin of BONCAT, ABPs were first focused on the enrichment and identification of the proteome and have only recently been applied for microbial identification. Genomic sequencing of ABP-enriched microbes is still in its infancy, with a limited number of published examples. However, with the specificity of ABPs towards phenotypes of interest, ABP–FACS-Seq offers a rich field to be explored.

An early application of ABP–FACS used probes designed to target β-glucuronidases in the mouse gastrointestinal microbiome. Hydrolysis by glucuronidases is of interest to the medical field for drug mechanisms and disease identification. In this study by Whidbey et al., an ABP was synthesized to mimic a glucuronidated metabolite leading to probe labeled populations containing 13 enriched OTUs compared to the bulk cell population (Whidbey et al. 2019). Interestingly, all 13 enriched OTUs were characterized as belonging to Bacillota, implicating members of this phylum as critical microbiota towards glucuronidase metabolism. Additionally, taxa at the family and genus levels were detected in both the probe-labeled and probe unlabeled sorted populations, further describing the dichotomy of activity in microbial populations and cell–cell heterogeneity of active phenotypes.

A strong benefit in the application of ABPs for microbial populations is that probes can be "pre-clicked" with a fluorophore prior to being introduced to viable cells (Steiger et al. 2020). This removes the need for microbes to be exposed to chemicals during the copper catalyzed click chemistry reaction. In one such study, researchers screened the soil microbiome for glycosidase activity while maintaining the viability of cells for post-FACS incubations (Reichart et al. 2023). By leaving cells unfixed and not exposed to click chemistry conditions, downstream of ABP–FACS samples were able to be incubated and showed enrichment of activity in probe-labelled populations compared to unlabeled or the bulk sorted populations. Phenotypes of interest (in this study, cellulolytic activity) from a complex matrix were queried using two separate ABPs to sort and enrich subpopulations from the native microbiome. While the application of ABPs for viable microbial cells is relatively recent, much promise is shown for the unique outcomes that can be achieved with this workflow. However, ABP–FACS may bias against any extracellular proteins as these proteins would be washed out of the sample during the preparation for FACS.

5.6 Outlook of Activity-Based Approaches

Outlined here are two approaches that originated to study the proteome but were later applied to interrogate the activity of microbial communities via bioorthogonal labeling, protein profiling, and flow cytometry. However, this is not an exhaustive review of potential methods for activity-labeling measurements. Additional work has been conducted using various labeled substrates (Doud et al. 2020), beads (Riva et al. 2023), and sensors (D'Angelo et al. 2021) to identify rare and unique microbial guilds responsible for specific phenotypes. Other molecules such as DNA (Beauchemin et al. 2023) or lipids (Lee et al. 2021) are also possible targets for activity-enrichment, as well as methods that utilize substrate incorporation to later be analyzed and enriched with other instrumentation, such as Raman-activated cell sorting (Zhang et al. 2015) or magnetic beads (Mena et al. 2022). Further comparisons would be needed to highlight the most applicable method for the hypotheses and sample types to be tested.

While this chapter focused on flow cytometry applications of activity studies with the goal of enriching whole cells, similar approaches have been described for enrichment of activity-associated proteins. In either situation, the resultant datasets provide a wealth of information for machine learning and modeling of microbial consortia. Using computational interrogation of phenotypic measurements describes an iterative approach of predictive phenomics to study microbial function, interactions, functional machinery, and networks that may further drive or prevent phenotypes in various systems, such as bioprocessing or viral infections. Methodological advancements to combine BONCAT, ABP, or other enriching activity approaches with FACS allow for more specific interactions to be studied on single-cell or small, pooled subsets of microbes. Coupling activity-based approaches with additional multi-omic measurements would build a more complete picture of microbial phenotypes and the functional machinery required for activity. As additional methods are developed for single-cell or low biomass input for transcriptomics (Kuchina et al. 2021) and proteomics (Fulcher et al. 2024), further characterization of BONCAT- or ABP–FACS-enriched populations, followed by multi-omics, may become possible.

Acknowledgments Thank you to Dr. Mária Džunková and Dr. Christian Rinke for the opportunity to provide insight into the field of activity-based studies. This work was supported by the Predictive Phenomics Initiative Laboratory Directed Research and Development Program. PNNL is operated by Battelle for the DOE under Contract DE-AC06-76RL01830.

References

Barglow KT, Cravatt BF (2007) Activity-based protein profiling for the functional annotation of enzymes. Nat Methods 4(10):822–827

Beauchemin ET, Hunter C, Maurice CF (2023) Actively replicating gut bacteria identified by 5-ethynyl-2′-deoxyuridine (EdU) click chemistry and cell sorting. Gut Microbes 15(1):2180317

Carini P, Marsden PJ, Leff JW, Morgan EE, Strickland MS, Fierer N (2016) Relic DNA is abundant in soil and obscures estimates of soil microbial diversity. Nat Microbiol 2:16242

Chauvigne-Hines LM, Anderson LN, Weaver HM, Brown JN, Koech PK, Nicora CD et al (2012) Suite of activity-based probes for cellulose-degrading enzymes. J Am Chem Soc 134(50):20521–20532

Couradeau E, Sasse J, Goudeau D, Nath N, Hazen TC, Bowen BP et al (2019) Probing the active fraction of soil microbiomes using BONCAT-FACS. Nat Commun 10(1):2770

Cravatt BF, Wright AT, Kozarich JW (2008) Activity-based protein profiling: from enzyme chemistry to proteomic chemistry. Annu Rev Biochem 77:383–414

D'Angelo T, Goordial J, Poulton NJ, Seyler L, Huber JA, Stepanauskas R, Orcutt BN (2021) Oceanic crustal fluid single cell genomics complements metagenomic and metatranscriptomic surveys with orders of magnitude less sample volume. Front Microbiol 12:738231

Dieterich DC, Link AJ, Graumann J, Tirrell DA, M. SE. (2006) Selective identification of newly synthesized proteins in mammalian cells using bioorthonogal noncanonical amino acid tagging (BONCAT). Proc Natl Acad Sci U S A 103(25):9482–9487

Doud DFR, Bowers RM, Schulz F, De Raad M, Deng K, Tarver A et al (2020) Function-driven single-cell genomics uncovers cellulose-degrading bacteria from the rare biosphere. ISME J 14(3):659–675
Du Z, Behrens SF (2021) Tracking de novo protein synthesis in the activated sludge microbiome using BONCAT-FACS. Water Res 205:117696
Fang H, Peng B, Ong SY, Wu Q, Li L, Yao SQ (2021) Recent advances in activity-based probes (ABPs) and affinity-based probes (AfBPs) for profiling of enzymes. Chem Sci 12(24):8288–8310
Frates ES, Spietz RL, Silverstein MR, Girguis P, Hatzenpichler R, Marlow JJ (2023) Natural and anthropogenic carbon input affect microbial activity in salt marsh sediment. Front Microbiol 14:1235906
Fulcher JM, Markillie LM, Mitchell HD, Williams SM, Engbrecht KM, Degnan DJ et al (2024) Parallel measurement of transcriptomes and proteomes from same single cells using nanodroplet splitting. Nat Commun 15(1):10614
Hatzenpichler R, Scheller S, Tavormina PL, Babin BM, Tirrell DA, Orphan VJ (2014) In situ visualization of newly synthesized proteins in environmental microbes using amino acid tagging and click chemistry. Environ Microbiol 16(8):2568–2590
Hatzenpichler R, Connon SA, Goudeau D, Malmstrom RR, Woyke T, Orphan VJ (2016) Visualizing in situ translational activity for identifying and sorting slow-growing archaeal-bacterial consortia. Proc Natl Acad Sci U S A 113(28):E4069–E4078
Hatzenpichler R, Krukenberg V, Spietz RL, Jay ZJ (2020) Next-generation physiology approaches to study microbiome function at single cell level. Nat Rev Microbiol 18(4):241–256
Hug LA (2024) The ever-changing tree of life. Nat Microbiol 9(8):1906–1908
Hug LA, Baker BJ, Anantharaman K, Brown CT, Probst AJ, Castelle CJ et al (2016) A new view of the tree of life. Nat Microbiol 1:16048
Jansson JK, Hofmockel KS (2018) The soil microbiome-from metagenomics to metaphenomics. Curr Opin Microbiol 43:162–168
Khalili B, Weihe C, Kimball S, Schmidt KT, Martiny JBH (2019) Optimization of a method to quantify soil bacterial abundance by flow cytometry. mSphere 4(5)
Kiick KL, Saxon E, Tirrell DA, Bertozzi CR (2001) Incorporation of azides into recombinant proteins for chemoselective modification by the Staudinger ligation. Proc Natl Acad Sci U S A 99(1):19–24
Krishnamoorthy S, Steiger AK, Nelson WC, Egbert RG, Wright AT (2022) An activity-based probe targeting the streptococcal virulence factor C5a peptidase. Chem Commun (Camb) 58(58):8113–8116
Krukenberg V, Reichart NJ, Spietz RL, Hatzenpichler R (2021) Microbial community response to polysaccharide amendment in anoxic hydrothermal sediments of the Guaymas Basin. Front Microbiol 12:763971
Kuchina A, Brettner LM, Paleologu L, Roco CM, Rosenberg AB, Carignano A et al (2021) Microbial single-cell RNA sequencing by split-pool barcoding. Science 371(6531)
Landor LAI, Bratbak G, Larsen A, Tjendra J, Vage S (2023) Differential toxicity of bioorthogonal non-canonical amino acids (BONCAT) in Escherichia coli. J Microbiol Methods 206:106679
Lee MT, Le HH, Johnson EL (2021) Dietary sphinganine is selectively assimilated by members of the mammalian gut microbiome. J Lipid Res 62:100034
Madill MBW, Luo Y, Sampara P, Ziels RM (2021) Activity-based cell sorting reveals resistance of functionally degenerate nitrospira during a press disturbance in nitrifying activated sludge. mSystems 6(4):e0071221
Marlow J, Spietz R, Kim KY, Ellisman M, Girguis P, Hatzenpichler R (2021) Spatially resolved correlative microscopy and microbial identification reveal dynamic depth- and mineral-dependent anabolic activity in salt marsh sediment. Environ Microbiol 23(8):4756–4777
McKay LJ, Smith HJ, Barnhart EP, Schweitzer HD, Malmstrom RR, Goudeau D, Fields MW (2022) Activity-based, genome-resolved metagenomics uncovers key populations and pathways involved in subsurface conversions of coal to methane. ISME J 16(4):915–926

Mena C, Reglero P, Balbin R, Martin M, Santiago R, Sintes E (2022) Dynamics of actively dividing prokaryotes in the western Mediterranean Sea. Sci Rep 12(1):2064
Meurer K, Barron J, Chenu C, Coucheney E, Fielding M, Hallett P et al (2020) A framework for modelling soil structure dynamics induced by biological activity. Glob Chang Biol 26(10):5382–5403
Michels DE, Lomenick B, Chou T-F, Sweredoski MJ, Pasulka A (2021) Amino acid analog induces stress response in marine Synechococcus. Appl Environ Microbiol 87(15):e0020021
Reichart NJ, Jay ZJ, Krukenberg V, Parker AE, Spietz RL, Hatzenpichler R (2020) Activity-based cell sorting reveals responses of uncultured archaea and bacteria to substrate amendment. ISME J 14(11):2851–2861
Reichart NJ, Steiger AK, Van Fossen EM, McClure R, Overkleeft HS, Wright AT (2023) Selection and enrichment of microbial species with an increased lignocellulolytic phenotype from a native soil microbiome by activity-based probing. ISME Commun 3(1):106
Reintjes G, Arnosti C, Fuchs B, Amann R (2019) Selfish, sharing and scavenging bacteria in the Atlantic Ocean: a biogeographical study of bacterial substrate utilisation. ISME J 13(5):1119–1132
Rinke C, Schwientek P, Sczyrba A, Ivanova NN, Anderson IJ, Cheng JF et al (2013) Insights into the phylogeny and coding potential of microbial dark matter. Nature 499(7459):431–437
Riva A, Rasoulimehrabani H, Cruz-Rubio JM, Schnorr SL, von Baeckmann C, Inan D et al (2023) Identification of inulin-responsive bacteria in the gut microbiota via multi-modal activity-based sorting. Nat Commun 14(1):8210
Schroder SP, de Boer C, McGregor NGS, Rowland RJ, Moroz O, Blagova E et al (2019) Dynamic and functional profiling of Xylan-degrading enzymes in aspergillus secretomes using activity-based probes. ACS Cent Sci 5(6):1067–1078
Schweitzer HD, Smith HJ, Barnhart EP, McKay LJ, Gerlach R, Cunningham AB et al (2022) Subsurface hydrocarbon degradation strategies in low- and high-sulfate coal seam communities identified with activity-based metagenomics. NPJ Biofilms Microbiomes 8(1):7
Steiger AK, Fansler SJ, Whidbey C, Miller CJ, Wright AT (2020) Probe-enabled approaches for function-dependent cell sorting and characterization of microbiome subpopulations. Methods Enzymol 638:89–107
Steward KF, Eilers B, Tripet B, Fuchs A, Dorle M, Rawle R et al (2020) Metabolic implications of using BioOrthogonal non-canonical amino acid tagging (BONCAT) for tracking protein synthesis. Front Microbiol 11:197
Taguer M, Shapiro BJ, Maurice CF (2021) Translational activity is uncoupled from nucleic acid content in bacterial cells of the human gut microbiota. Gut Microbes 13(1):1–15
Trexler RV, Van Goethem MW, Goudeau D, Nath N, Malmstrom RR, Northen TR, Couradeau E (2023) BONCAT-FACS-Seq reveals the active fraction of a biocrust community undergoing a wet-up event. Front Microbiol 14:1176751
Valentini TD, Lucas SK, Binder KA, Cameron LC, Motl JA, Dunitz JM, Hunter RC (2020) Bioorthogonal non-canonical amino acid tagging reveals translationally active subpopulations of the cystic fibrosis lung microbiota. Nat Commun 11(1):2287
Vereecken H, Schnepf A, Hopmans JW, Javaux M, Or D, Roose T et al (2016) Modeling soil processes: review, key challenges, and new perspectives. Vadose Zone J 15(5):1–57
Webber LC, Anderson LN, Paraiso IL, Metz TO, Bradley R, Stevens JF, Wright AT (2023) Affinity- and activity-based probes synthesized from structurally diverse hops-derived xanthohumol flavonoids reveal highly varied protein profiling in Escherichia coli. RSC Adv 13(42):29324–29331
Whidbey C, Sadler NC, Nair RN, Volk RF, DeLeon AJ, Bramer LM et al (2019) A probe-enabled approach for the selective isolation and characterization of functionally active subpopulations in the gut microbiome. J Am Chem Soc 141(1):42–47
Wilhelm RC, Singh R, Eltis LD, Mohn WW (2019) Bacterial contributions to delignification and lignocellulose degradation in forest soils with metagenomic and quantitative stable isotope probing. ISME J 13(2):413–429

Willems LI, Van Der Linden WA, Li N, Li K-Y, Liu N, Hoogendoorn S et al (2011) Bioorthogonal chemistry: application in activity-based protein profiling. Acc Chem Res 44(9):718–729

Zegeye EK, Sadler NC, Lomas GX, Attah IK, Jansson JK, Hofmockel KS et al (2021) Activity-based protein profiling of chitin catabolism. Chembiochem 22(4):717–723

Zhang Q, Zhang P, Gou H, Mou C, Huang WE, Yang M et al (2015) Towards high-throughput microfluidic Raman-activated cell sorting. Analyst 140(18):6163–6174

Technical Advancements in Microbial Single-Cell Omics Analysis

6

Haruko Takeyama, Masahito Hosokawa, Masahiro Ando, and Yohei Nishikawa

Abstract

Recent advancements in single-cell omics technologies have significantly enhanced our ability to analyze individual microbial cells at the genomic, transcriptomic, and metabolomic levels. While single-cell analysis has traditionally focused on eukaryotic systems, emerging methodologies now enable high-throughput studies of environmental microbiomes and viruses, providing deeper insights into microbial ecology and functional diversity.

Microfluidic droplet-based single-cell genomics has improved whole-genome amplification efficiency while minimizing contamination, enabling the sequencing of thousands of microbial genomes with higher accuracy. Targeted single-cell genomics techniques have further refined genome recovery by selectively isolating cells containing specific genetic sequences. These innovations have advanced our understanding of microbial community structures and host-microbe interactions.

H. Takeyama (✉) · M. Hosokawa
Department of Life Science and Medical Bioscience, Waseda University, Tokyo, Japan

Research Organization for Nano & Life Innovation, Waseda University, Tokyo, Japan

Institute for Advanced Research of Biosystem Dynamics, Waseda Research Institute for Science and Engineering, Graduate School of Advanced Science and Engineering, Waseda University, Tokyo, Japan
e-mail: haruko-takeyama@waseda.jp

M. Ando
Research Organization for Nano & Life Innovation, Waseda University, Tokyo, Japan

Y. Nishikawa
Research Organization for Nano & Life Innovation, Waseda University, Tokyo, Japan

Biomanufacturing Process Research Center, National Institute of Advanced Industrial Science and Technology (AIST), Ibaraki, Japan

C. Rinke, M. Džunková (eds.), *Single-Cell Omics in Microbiome Research*, The Microbiomes of Humans, Animals, Plants, and the Environment 7,
https://doi.org/10.1007/978-3-032-07527-7_6

Single-cell transcriptomics have revealed substantial gene expression variability within bacterial populations, highlighting mechanisms of stress adaptation, antibiotic resistance, and virulence. Despite technical challenges, including the lack of poly-A tails and low RNA content, recent breakthroughs in scRNA-seq methodologies have improved sensitivity and throughput, enabling high-resolution transcriptional profiling.

Single-cell metabolomics, particularly through Raman microspectroscopy, offers label-free, nondestructive metabolic profiling at subcellular resolution. Integration with computational tools such as multivariate curve resolution (MCR) analysis has facilitated the identification of metabolic heterogeneity and cellular differentiation processes. These approaches are increasingly being applied to biotechnology, environmental microbiology, and biomedical research.

Future advancements in single-cell omics, including multi-omics integration and AI-driven data analysis, will further enhance our understanding of microbial functions and interactions. These technologies hold promises for diverse applications, from microbial ecology to biomanufacturing, by enabling unprecedented insights into the molecular mechanisms that govern microbial life.

Keywords

Single-cell analysis · Multi-omics · Droplet microfluidics · Raman-microspectroscopy

6.1 Introduction

Recent advancements in data acquisition technologies have increasingly enabled the extraction of detailed information from trace biological components within individual microbial cells. Simultaneously, the comprehensive understanding of complex biological phenomena demands a multi-omics approach, which integrates diverse perspectives to uncover intricate cellular processes. To address this need, significant efforts have been directed toward the development of advanced methodologies and technologies.

Historically, the field of single-cell analysis has primarily focused on eukaryotic cells, with microbial studies lagging behind in technological progress. Despite this gap, recent improvements in detection sensitivity and resolution have paved the way for emerging applications of single-cell omics analysis in microbial systems. These advances are beginning to reveal unprecedented insights into the genetic, transcriptomic, and metabolic landscapes of individual microbial cells, providing a deeper understanding of their roles in complex environments.

This introduction highlights the latest innovations in high-throughput and high-precision single-cell omics technologies, with a particular focus on their applications to genome, transcriptome, and metabolome analyses of environmental microbes. These developments represent a critical step forward in leveraging single-cell methodologies to unravel the complexity of microbial life at an unprecedented level of detail.

6.2 Microfluidic Droplet-Based Single-Cell Genomics of Environmental Microbiome and Viruses

6.2.1 Microfluidic Droplet-Based Single-Cell Genomics of Environmental Microbiome

The quantity of DNA contained in a single microbial cell is on the order of femtograms (10^{-15} g), which is about 1000 times less than that of eukaryotic cells. These amounts do not meet the current requirements for high-throughput sequencing, therefore necessitating a whole-genome amplification (WGA) step. So far, several efforts have been made to improve WGA methods, including polymerase development through in vitro protein evolution and the optimization of reaction mixture compositions (Oyola et al. 2014; Zhang et al. 2023). Nevertheless, particularly in the amplification of minute amounts of microbial DNA, the quality of the amplified DNA is compromised by amplification biases and the amplification of nontarget (contaminating) DNA, which is a bottleneck in single-cell genomics (Hosokawa and Nishikawa 2024).

As a possible solution to these problems, a method using nanoliter-scale reaction chambers for WGA has been proposed (Marcy et al. 2007). By reducing the reaction volume, the risk of contaminating DNA is suppressed, and the amplification bias is mitigated by preventing excessive DNA amplification. Subsequently, several methods have been developed for WGA in nanoliter-scale reaction chambers on devices or substrates (Leung et al. 2016, 2012; Xu et al. 2016). These techniques improved the ease of handling, sequence accuracy, and the number of cells that can be processed. Along with these techniques, methods have been reported using picoliter-scale droplets created with microfluidic devices as reaction chambers (Nishikawa et al. 2015; Rhee et al. 2016). Droplets are generated by shearing a cell suspension with carrier oil containing a surfactant. Microfluidic devices enable the generation of monodispersed and picoliter-volume reaction vessels in a continuous and high-throughput manner. These characteristics have been applied as a tool for single-cell analysis of various cell types, including microbiomes.

Early reports demonstrated that encapsulating DNA fragments extracted from a single cell into individual droplets for amplification could suppress amplification bias. After that, methods were developed to encapsulate single cells into individual droplets for parallel WGA, significantly improving the throughput of single-cell genomics (Hosokawa et al. 2017; Lan et al. 2016). So far, several groups reported microfluidic droplet-based single-cell WGA methods. In the methods named single-cell genomic sequencing (SiC-seq) (Lan et al. 2017) or Microbe-seq (Zheng et al. 2022), single cells are encapsulated within droplets, and after the cell lysis library preparation is carried out within the droplets either directly or after DNA amplification (Fig. 6.1a). These methods allow the acquisition of genome sequences from

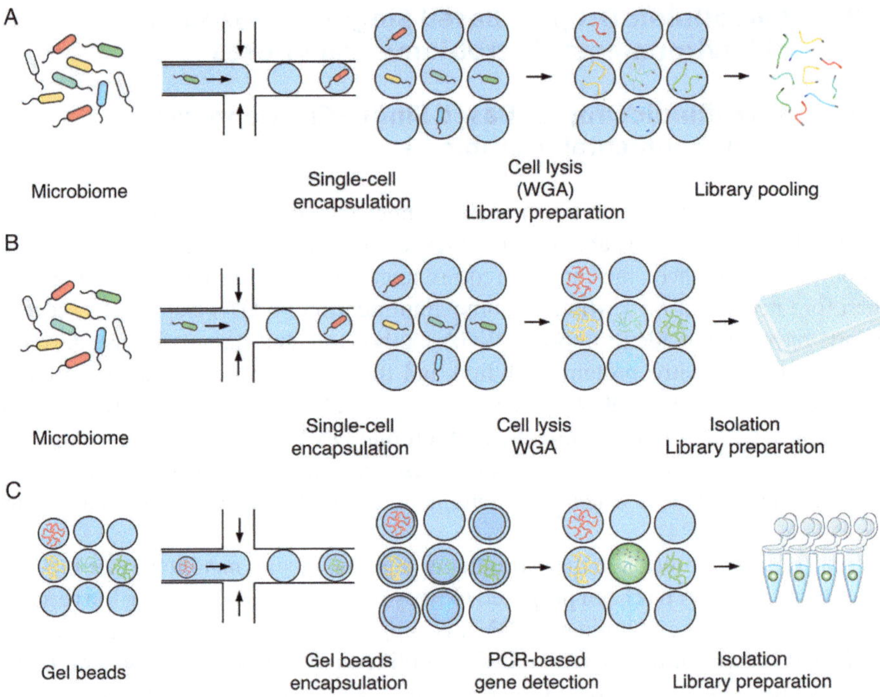

Fig. 6.1 Microfluidic droplet-based single-cell genomics. (**a**) The method where all steps from single-cell encapsulation to library preparation are completed within droplets. Mechanisms for adding reagents, such as droplet fusion, are required to achieve multiple reaction steps within the droplet. (**b**) The method to complete the steps from single-cell encapsulation to whole genome amplification within droplets. Droplets containing a gelling agent (gel beads) are used to capture cells and high molecular weight DNA, and multiple reaction steps can be easily demonstrated by changing the reagent in the dispersed phase. Library preparation after gel bead sorting into multi-well plates requires a high-throughput handling system. (**c**) The method for targeted single-cell genomics. To selectively analyze droplets with specific sequences, gel beads with amplified DNA are suspended in PCR reagents for gene detection and introduced into the microfluidic device. By isolating and sequencing droplets in which the target gene has been detected, it is possible to selectively analyze the genomic sequences of cells with specific sequences. (Figure created with BioRender)

tens of thousands of single cells at a single high-throughput sequencing. However, the genome coverage per single cell is relatively limited, and the completeness of SAGs can be improved by in silico integration at the species level; i.e., combining the genomic information from multiple microbes derived from distinct droplets. In addition, equipment for droplet fusion is required to complete the entire process within droplets. On the other hand, in the method named single-cell amplified genome in gel (SAG-gel) (Chijiiwa et al. 2020; Nishikawa et al. 2022), the WGA process is carried out within agarose-containing droplets (gel beads) (Fig. 6.1b). The droplets with amplified DNA are then sorted by fluorescence-activated cell sorting (FACS) into multi-well plates for plate-based barcoding. This method can obtain thousands of SAGs at a single high-throughput sequencing, achieving relatively higher genome coverage from a single cell. However, robotics technologies such as automated pipetting machines are recommended to be introduced for high-throughput genome sequencing.

The development of single-cell genomics technologies using microfluidic droplets has suppressed the amplification bias and the risk of DNA contamination, contributing to the accumulation of qualified single-amplified genomes (SAGs). The increased number of SAGs constructed in a single high-throughput sequencing has made it possible to obtain sequences from diverse microbial lineages in environmental samples. Moreover, integrating multiple SAGs derived from the same species or strain can improve SAG completeness, while a bioinformatics tool ccSAG has been developed to enable highly accurate sequence integration by cross-referencing the sequences of each SAG (Kogawa et al. 2018). On the other hand, random sorting and sequencing of amplified DNA within gel beads have often resulted in excessive duplication of SAGs derived from the same species or strain, which can be a waste of sequencing effort. To address this issue, new methods have been proposed to isolate droplets containing target sequences selectively.

In a further developed version of the SAG-gel method, the gel beads are suspended in a PCR reaction mixture containing primers and probes to detect target genes after performing single-cell WGA within them. Then, the gel bead suspension is introduced into a microfluidic device to generate droplets encapsulating the gel beads. The probes label the droplets containing the target sequence by performing PCR on the collected droplets. Targeted single-cell genomics is achieved by selecting and sequencing the droplets that show the probe signal (Ide et al. 2022) (Fig. 6.1c). Another method involves generating droplets with cells suspended in a PCR reaction mixture to detect target sequences. After the PCR reaction, droplets are introduced into the sorting device and deflected by dielectrophoresis into the collection channel. Then, the contents of the sorted droplets are recovered for whole-genome amplification to collect the genome sequences of target microbiomes (Pryszlak et al. 2022). These methods have been demonstrated to apply to obtaining sequences of specific strains from diverse microbial populations and samples with a high risk of DNA contamination, such as host symbiotic bacteria.

6.2.2 Microfluidic Droplet-Based Single-Cell Genomics of Environmental DNA Viruses

The single-cell genomics platform previously described for environmental microbiomes has begun to be applied to single-particle genome sequencing for environmental DNA viruses. Viruses capable of infecting environmental microbes (mainly bacteriophages that infect bacteria) play a key role in shaping microbial communities by reducing host population sizes and facilitating gene transfer among prokaryotes. In addition, these viruses are known to have greater genome sequence diversity than prokaryotes. Therefore, analyzing interactions between prokaryotes and viruses is necessary for further understanding the ecology of environmental microbiomes. The genome sequencing method for single-particle viruses was first reported in 2011 (Allen et al. 2011). The method involved viral particle sorting with a cell sorter after DNA staining with fluorescence dye, followed by WGA and library preparation. While genome analysis techniques at the single-virus particle level have attracted attention (Martinez-Hernandez et al. 2017; Martinez Martinez et al. 2020), sorting viral particles with tiny sizes and DNA amounts using a cell sorter has remained technically challenging, resulting in low throughput and accuracy. To overcome this problem, a method for single-particle genomics of DNA viruses was proposed to modify the SAG-gel platform (Nishikawa et al. 2024). This method involves amplifying viral genomes within gel beads and subsequent sorting with FACS. This method has been reported to enable the recovery of various viral-like contigs from river water samples, including single-stranded DNA viruses. It was also possible to detect genomic microdiversity within the same viral species, contributing to the elucidation of part of the host adaptation strategy of the virus. Alongside the development of genome sequencing methods, bioinformatics tools specialized for analyzing viral sequences are also being developed. A recently reported tool called vClean can assess the risk of contamination of heterologous viral sequences against sequences constructed by metavirome and single-virus genomics (Wagatsuma et al. 2025). When a viral sequence shows a high contamination risk, only the longest contig is adopted, thereby preventing the accumulation of incorrect viral sequences in the database.

6.2.3 Future Perspectives

Microfluidic droplet-based single-cell genomics is adaptable to various environments and a wide range of cell types, which enables the accumulation of unexplored genetic resources. These sequence data are expected to significantly contribute to the field of biomanufacturing and engineering biology by linking with AI-based protein structure prediction technology and genetic engineering technology. Also, large-scale sequencing of viral genomes will improve our understanding of microbe-virus interactions and enable us to capture the gene transfer in the environment at a higher resolution. Sequence data of unknown environmental viruses may provide insight into how viruses modulate the microflora. In recent years, phage therapy, in which lysogenic phages are used to combat harmful bacteria, such as antibiotic-resistant bacteria, has received renewed attention. The accumulation of novel phage sequences will help to expand the potential of phages as a tool to control the microflora.

The primary advantages of microfluidic droplets are the high-throughput compartmentalization of single cells and the ability to provide compartmentalized spaces as reaction vessels for detecting biological components. Therefore, not only microbiomes but also viruses and membrane vesicles can be targeted for analysis. By combining the composition of droplets and the detection mechanism for biological components, this technology is also expected to have the potential to contribute to analyses other than single-cell genomics.

6.3 Single-Cell Transcriptomics of Microbes

Bacterial populations demonstrate significant gene expression variability, even among isogenic groups, which plays critical roles in stress responses (Sampaio et al. 2022), antibiotic resistance (Arnoldini et al. 2014; Helaine et al. 2014), and virulence (Hockenberry et al. 2021). Traditional bulk RNA sequencing averages these expressions across populations, obscuring individual variability. Single-cell RNA sequencing (scRNA-seq) addresses this limitation by offering insights into cellular heterogeneity.

The journey of scRNA-seq began with eukaryotes in 2009 (Tang et al. 2009), enabling unprecedented analysis of cellular heterogeneity. However, applying this technology to bacterial systems has proven more challenging due to their distinct biological characteristics. Early bacterial scRNA-seq efforts addressed these unique hurdles, paving the way for current state-of-the-art methodologies (Brennan and Rosenthal 2021; Homberger et al. 2022; Imdahl and Saliba 2020). Advances such as split-pool barcoding and droplet microfluidics have been instrumental in pushing the field forward.

6.3.1 Technical Challenges

Bacterial scRNA-seq faces several significant technical challenges due to the inherent characteristics of bacteria. A primary challenge is the absence of poly-A tails in bacterial mRNAs, which differs from eukaryotic mRNAs. This absence necessitates the development of alternative enrichment strategies, such as not-so-random hexamer priming (Hayashi et al. 2018; Nishimura et al. 2023) or custom capture probes (McNulty et al. 2023).

The predominance of ribosomal RNA in bacterial cells presents another substantial hurdle. To effectively focus on mRNAs, researchers have developed various depletion techniques, including RNase H (Ma et al. 2023; Wang et al. 2023a) and Cas9-based methods (Gu et al. 2016; Homberger et al. 2023; Nishimura et al. 2023; Prezza et al. 2020; Shen et al. 2024; Xu et al. 2023). The limited RNA content in bacterial cells compounds these challenges, requiring exceptionally sensitive capture and amplification technologies. These include the use of SuperScript IV reverse transcriptase (Homberger et al. 2023; Zucha et al. 2020) and unique molecular identifiers (Kivioja et al. 2011) to minimize biases in the amplification process.

Cell wall diversity among bacterial species creates additional complexity, as it necessitates the development of species-specific lysis protocols to extract RNA while maintaining its integrity. The structure of bacterial operons presents its own set of challenges, specifically requiring methods capable of capturing full-length transcripts. Furthermore, the accurate detection of noncoding RNAs, particularly short RNAs (sRNAs), introduces additional technical difficulties due to their limited length.

These diverse technical challenges collectively emphasize the necessity for specialized tools and protocols specifically designed for bacterial systems. The field continues to evolve as researchers develop innovative solutions to address these unique characteristics of bacterial RNA analysis at the single-cell level.

6.3.2 Methodological Approaches

Bacterial scRNA-seq methodology has evolved to encompass three distinct yet complementary approaches, each designed to address specific research needs and technical challenges (Fig. 6.2). The first approach centers on fluorescence-activated cell sorting (FACS)-based methods, exemplified by techniques such as MATQ-seq (Homberger et al. 2022; Imdahl et al. 2020) and RamDA-seq (Nishimura et al. 2023). Among all available bacterial scRNA-seq methods to date, RamDA-seq has achieved unprecedented sensitivity and demonstrated the highest gene detection capability. This method not only provides near full-length transcript coverage but also enables comprehensive operon structure analysis, representing a significant advancement in the field of bacterial transcriptomics. While FACS-based methods may be limited in terms of throughput, their strength lies in their ability to detect low-abundance transcripts and sRNAs, with RamDA-seq setting the current standard for sensitivity and detection capability.

The second methodological approach involves split-pool barcoding, with notable techniques including PETRI-seq (Blattman et al. 2020) and microSPLiT (Kuchina et al. 2021). These methods represent a significant advancement in high-throughput analysis, enabling researchers to examine tens of thousands of cells simultaneously. While they may sacrifice some sensitivity compared to FACS-based approaches, their high-throughput capabilities make them particularly valuable for large-scale studies investigating population dynamics.

The third approach utilizes droplet-based methods, such as BacDrop (Ma et al. 2023) and smRandom-seq (Shen et al. 2024; Xu et al. 2023). These techniques strike a balance between throughput and sensitivity by encapsulating cells with barcoded beads in droplets. This approach has proven particularly effective for efficient profiling of large bacterial populations, with BacDrop (Ma et al. 2023) specifically demonstrating its utility in uncovering heterogeneity in antibiotic responses within bacterial communities.

The diversity of these methodological approaches highlights the importance of careful method selection based on specific research objectives. Each approach presents its own set of advantages and limitations in terms of sensitivity, throughput, and application scope, allowing researchers to choose the most appropriate method for their particular research questions and experimental constraints.

6.3.3 Applications and Insights

Bacterial scRNA-seq has revolutionized microbial biology research by providing unprecedented insights into cellular heterogeneity and functional diversity. Investigations into transcriptional heterogeneity have revealed surprising variability within isogenic populations, uncovering the existence of stress-response subpopulations and bet-hedging strategies. A notable example comes from M3-seq (Wang et al. 2023a), which identified distinct acid-tolerant *Escherichia coli* subpopulations

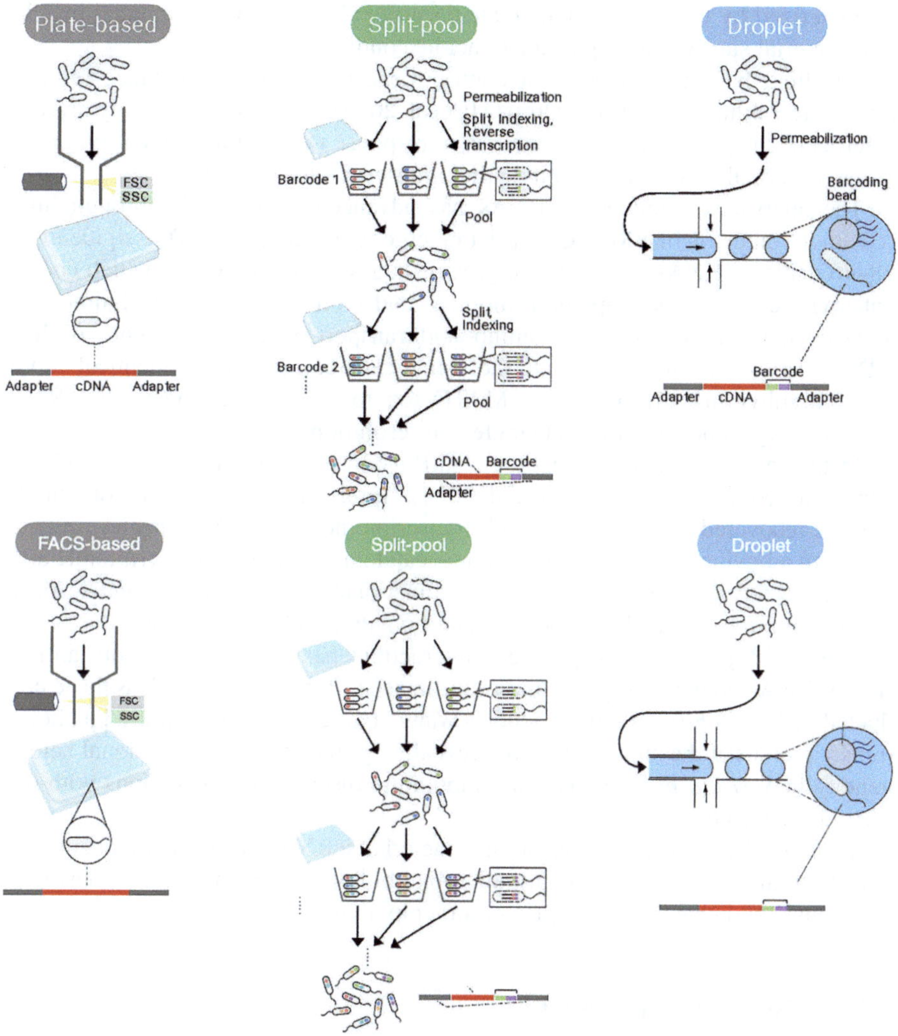

Fig. 6.2 Overview of three approaches for bacterial single-cell RNA sequencing. FACS-based methods (left) isolate single cells by fluorescence-activated cell sorting into individual wells, enabling high-sensitivity detection of transcripts, including low-abundance RNAs, but with limited throughput. Split-pool methods (middle) distribute cells across multiple wells with unique barcoding in sequential rounds, allowing high-throughput analysis of thousands of cells simultaneously with reduced handling. Droplet-based methods (right) encapsulate cells with barcoded beads in microfluidic droplets, providing balanced throughput and sensitivity by creating isolated reaction environments for single cells. (Figure created with BioRender)

characterized by upregulated glutamate decarboxylases, providing crucial insights into bacterial survival strategies under acidic conditions.

The field has made significant contributions to antibiotic-resistance research through techniques such as BacDrop (Ma et al. 2023). BacDrop specifically revealed heterogeneous gene-expression patterns in response to antibiotics, identifying persister cell populations that exhibit distinct transcriptional signatures.

The analysis of noncoding RNAs has advanced significantly through high-sensitivity methods like MATQ-seq (Homberger et al. 2023). MATQ-seq identified 28–46 distinct sRNAs at successive growth stages, including CsrB, which coordinates virulence, stress adaptation, motility, and biofilm formation in Salmonella, and GcvB, a central regulator of amino-acid transport systems. Expression of these sRNAs varies substantially among individual cells cultured under identical conditions, highlighting the ability of MATQ-seq to resolve heterogeneous sRNA-mediated regulatory programs at single-cell resolution.

In the realm of microbial ecology, smRandom-seq has proven invaluable, as demonstrated in recent work (Jia et al. 2024). Applied to the rumen microbiota, the method recovered 174,531 single-cell transcriptomes from 2534 species; integration with a bovine gastrointestinal genome catalogue yielded a transcriptomic atlas that delineated 172 core active taxa in 12 functional clusters and pinpointed *Basfia succiniciproducens* as a keystone of the carbohydrate-metabolic niche.

The investigation of host-pathogen interactions has also benefited substantially from scRNA-seq technologies. Studies utilizing smRandom-seq, such as those conducted in *Drosophila* infection models (Wang et al. 2024), have captured pathogen transcriptomes at the single-cell level, revealing pronounced transcriptional heterogeneity in *Serratia marcescens* and pinpointing discrete subpopulations with distinct virulence potential.

Through these various applications, bacterial scRNA-seq has demonstrated its versatility and power in advancing microbial biology across multiple scales, from individual cell behavior to complex community dynamics.

6.3.4 Future Perspectives

The future of bacterial scRNA-seq centers on several key developments in both technical capabilities and applications. Enhanced sensitivity in detection methods will improve our ability to analyze low-abundance transcripts and noncoding RNAs, potentially uncovering subtle gene regulatory networks that are currently difficult to detect. Increased throughput capabilities, scaling up to millions of cells per experiment, will enable a more robust analysis of rare cell states and complex microbial communities, which is particularly beneficial for environmental microbiome studies.

Expanding these techniques to less-studied bacteria, including unculturable species, will enhance our understanding of microbial diversity and function, potentially revealing novel metabolic pathways and ecological roles. The integration of multi-omics approaches with scRNA-seq shows promise for providing holistic insights into bacterial cell states, while computational innovation will be essential for addressing current challenges in data analysis through sophisticated workflows that can integrate metagenomics and single-cell data.

Through its evolution, bacterial scRNA-seq has proven to be a transformative tool in microbiology, successfully overcoming significant technical hurdles to unveil unprecedented insights into bacterial heterogeneity and gene regulation. By capturing cellular diversity at an unparalleled resolution, this technology has enabled researchers to decode complex phenomena such as microbial community interactions and host-pathogen dynamics. As these methodological innovations continue to advance, bacterial scRNA-seq promises to further revolutionize our understanding of microbial biology, unlocking new opportunities for discovery in microbiology and beyond.

6.4 Single-Cell Metabolomics of Microbes

6.4.1 Single-Cell Metabolomics

With rapid technological advancements, single-cell metabolomics has emerged as a cutting-edge frontier in molecular biology, which provides crucial insights into cellular function and heterogeneity at the individual cell level (Taylor et al. 2021; Wang et al. 2023b). By directly reflecting cell functions, metabolites offer a unique window into the biochemical processes occurring within cells (Liu and Locasale 2017; Nicholson et al. 1999). This approach complements other single-cell omics technologies by providing real-time snapshots of cellular activity, enabling researchers to uncover fundamental biological mechanisms and develop novel applications across multiple fields, from basic cell biology to biotechnology.

Recent advances in mass spectrometry (MS)-based analytical techniques have revolutionized single-cell metabolomics in eukaryotic cells, particularly through the integration of electrospray ionization (ESI)-MS, matrix-assisted laser desorption/ionization (MALDI)-MS, and secondary ion mass spectrometry (SIMS) (Liu and Yang 2021; Qin et al. 2023). The implementation of nano-electrospray ionization mass spectrometry (nanoESI-MS) coupled with integrated microfluidic platforms and flow cytometry has significantly enhanced analytical sensitivity and molecular specificity at the single-cell level (Feng et al. 2015; Taylor et al. 2021). Despite persistent technical challenges, such as ultralow analyte concentrations and minimal sample volumes inherent to single-cell analyses, these advanced methodologies now enable simultaneous quantification and characterization of hundreds of distinct metabolites within individual eukaryotic cells, offering unprecedented insights into cellular heterogeneity and metabolic variations. This technological evolution has opened new avenues for applications in precision medicine, particularly in understanding cellular metabolism and therapeutic responses in complex eukaryotic systems (Dakal et al. 2024; Li et al. 2023; Qin et al. 2023).

Building on these advances, efforts to develop single-cell metabolomics in microorganisms are progressing through the integration of microfluidic devices with mass spectrometry (Schirmer and Dusny 2023). This integration enables precise control over cell manipulation and metabolite analysis at microscale levels, overcoming limitations of traditional bulk analyses. A significant advance has been made through the development of droplet microfluidic technology for cell encapsulation, combined with advanced MS technologies such as Fourier transform ion cyclotron resonance mass spectrometry (FT-ICR-MS) and ESI-MS. These technologies now allow tracking of metabolites in small cell populations (15–20 cells), revealing cellular heterogeneity and environmental responses previously masked in population-averaged data (Dusny et al. 2019; Schirmer et al. 2020).

Despite these remarkable advances in MS-based metabolomics, achieving comprehensive single-cell resolution analysis remains a significant challenge, particularly in microorganisms. As described above, current approaches typically require pooling of several dozens of cells to achieve detectable metabolite levels—a fundamental limitation, as metabolites, unlike nucleic acids, cannot be amplified. This analytical constraint is most pronounced in microorganisms whose cell volumes (femtoliters) are approximately 1/1000 that of mammalian cells. The combination of minimal cell volumes, inability to amplify metabolites, and inevitable sample loss during preparation presents significant technical hurdles for absolute single-cell analysis. The development of more sensitive technologies will be crucial for advancing toward genuine single-cell metabolomic profiling, particularly for understanding transient metabolic states and rare cell populations.

6.4.2 Raman-Based Approach for Microbial Single-Cell Metabolomics

Complementing mass spectrometric approaches, Raman microspectroscopy has emerged as a powerful analytical platform that provides label-free, high-spatiotemporal resolution imaging of metabolites in living cells (Du et al. 2020; Jayan et al. 2022; Yue and Cheng 2016). Confocal Raman microspectroscopy offers unique capabilities for nondestructive, label-free metabolite analysis at single-cell resolution by combining Raman spectroscopy with confocal microscopy (Puppels et al. 1990). In this technique, laser light focused through a microscope excites molecular vibrations in the sample, and the resulting Raman scattered light is collected and analyzed to generate molecular fingerprint spectra (Pezzotti 2021). The confocal optical arrangement enables sub-μm spatial resolution, allowing for detailed molecular mapping within individual cells (Fig. 6.3). This approach shares operational similarities with confocal fluorescence microscopy, making it relatively accessible to researchers familiar with biological imaging techniques.

The application of Raman microspectroscopy to single-cell metabolomics faces several technical and analytical challenges. The inherent molecular complexity of cellular systems generates overlapping spectral signatures, complicating the isolation and quantification of individual metabolites. While modern analytical approaches incorporating statistical analysis and machine learning can differentiate

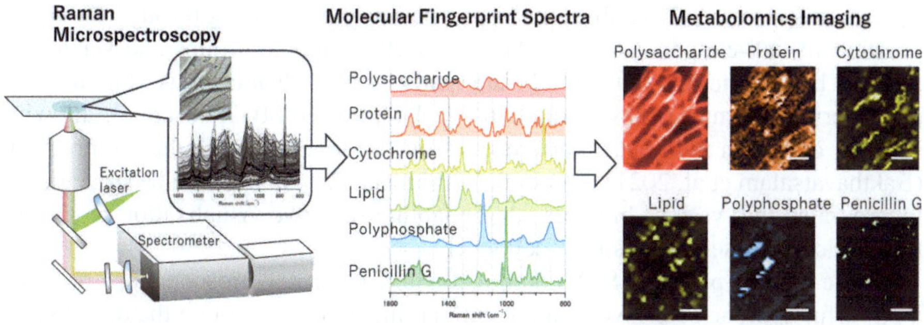

Fig. 6.3 Microbial single-cell metabolomics imaging by Raman microspectroscopy. Scale bars in mapping images are 5 μm

between cell populations, these methods often lack clear mechanistic interpretation of the underlying molecular changes. Additionally, the extended acquisition times necessary for three-dimensional mapping constrain the technique's capacity to capture dynamic metabolic processes. Advancing beyond these limitations requires sophisticated analytical solutions that can extract detailed molecular information while preserving the noninvasive advantages of Raman spectroscopy for single-cell analysis.

To address these complex analytical challenges, several advanced Raman spectroscopic techniques have been developed (Cui et al. 2022). These methodological innovations aim to enhance molecular specificity and detection sensitivity while maintaining the nondestructive nature of Raman spectroscopy. Four key approaches have shown particular promise in advancing single-cell metabolomics analysis: nonlinear Raman spectroscopy for enhanced sensitivity and speed, Raman tag labeling for specific molecular targeting, isotope-labeled Raman spectroscopy for metabolic pathway analysis, and computational spectral decomposition methods for label-free molecular identification. Each of these techniques offers distinct advantages in resolving the spectral complexity inherent to biological samples, with computational approaches complementing the enhanced molecular specificity achieved through physical and chemical methods at the single-cell level.

Nonlinear Raman techniques, including coherent anti-stokes Raman scattering (CARS) and stimulated Raman scattering (SRS), have emerged as powerful tools for high-sensitivity molecular imaging (Yue and Cheng 2016). CARS utilizes the interaction of three laser pulses to generate coherent signals from specific molecular vibrations, while SRS employs two laser pulses to excite molecular vibrations through stimulated excitation. These techniques enable high-speed, label-free visualization of biomolecular distributions and dynamics in living cells, including lipids, proteins, and DNA, with enhanced sensitivity compared to spontaneous Raman scattering. The application of these advanced imaging modalities has provided new insights into cellular processes, such as the dynamics of lipid droplets and alterations in lipid metabolism.

Raman spectroscopy enables metabolic pathway monitoring through the incorporation of cell-exogenous molecular tags such as alkyne groups and stable isotopes. The alkyne tags ($C \equiv C$) are particularly advantageous due to their characteristic Raman peaks in the cell-silent region (1800–2800 cm^{-1}), allowing specific detection without interference from endogenous cellular components (Bakthavatsalam et al. 2021). These bio-orthogonal Raman tags can be strategically incorporated into various biomolecules, enabling selective visualization of specific metabolic processes at the single-cell level.

Stable isotope probing (SIP) combined with Raman spectroscopy also provides a powerful approach for investigating cellular metabolism through the detection of spectral shifts induced by isotope incorporation (Cui et al. 2022; Xu et al. 2021). This technique typically employs isotopes such as ^{13}C, ^{15}N, and ^{2}H, which generate distinctive spectral signatures upon metabolic incorporation. For instance, deuterium incorporation from deuterated substrates results in characteristic C-D Raman bands (2000–2300 cm^{-1}), enabling both qualitative and semi-quantitative assessment of metabolic activity through the measurement of C-D band intensity and (C-D)/((C-H) + (C-D)) intensity ratios. This methodology has been successfully applied to investigate various cellular processes, including carbon and nitrogen fixation, resuscitation of viable but nonculturable (VBNC) cells, and antimicrobial susceptibility (Wang et al. 2016).

Taking Raman-based microbial analysis a step further, optofluidic Raman-activated cell sorting (RACS) has emerged as an important technique for isolating cells based on their Raman spectral signatures (Tang et al. 2024). This method integrates Raman microspectroscopy with microfluidic sorting systems, allowing for the selective isolation of cells with specific metabolic phenotypes detected through their distinctive Raman spectra. RACS provides a cultivation-independent approach to separate functionally active cells from complex microbial communities, with sorted cells remaining viable for subsequent genomic analysis (Lee et al. 2021b; Kogawa et al. 2022). This capability creates a crucial link between cellular metabolic functions and their genetic basis, enabling targeted genomic studies of microorganisms selected for their specific metabolic activities.

Meanwhile, recent advances in computational analysis have revolutionized the molecular discrimination capabilities of Raman microspectroscopy. Notably, multivariate curve resolution (MCR) analysis, which operates under physicochemically reasonable constraints, enables decomposition of complex cellular Raman spectra into distinct molecular components (Ando and Hamaguchi 2014; Tauler 1995). This mathematical approach can effectively separate overlapping spectral signatures and generate quantitative spatial distribution maps of specific metabolites within single cells. The integration of MCR with high-resolution Raman imaging has transformed metabolite analysis from simple classification tasks to comprehensive molecular profiling, providing detailed insights into metabolic dynamics and subcellular organization (Ando et al. 2023).

6.4.3 Application of Raman Metabolomics and MCR Analysis

The integration of Raman microspectroscopy with MCR analysis has enabled high-resolution, nondestructive metabolomics imaging at the single-cell level in microbial systems. This technical advancement allows for direct observation of metabolic dynamics in individual microbial cells during their differentiation processes, providing unprecedented insights into the spatiotemporal regulation of secondary metabolism (Horii et al. 2023). The methodology's nondestructive nature and high spatial resolution make it particularly valuable for understanding the relationship between cellular development and metabolite production in microorganisms.

High-resolution studies at the single-cell level have revealed intricate patterns of metabolite distribution within microbial cellular structures, demonstrating the power of this analytical approach. Studies of *Streptomyces avermitilis* have shown selective metabolite accumulation in specific morphological features during differentiation (Horii et al. 2023), while analysis of *Penicillium chrysogenum* has revealed heterogeneous, particle-like distributions of secondary metabolites within individual cells (Horii et al. 2020). These observations have provided direct evidence for specific transport mechanisms and subcellular localization patterns in these important microbial species (Samuel et al. 2022).

The technique's applications extend beyond single-cell analysis to microbial colony-level studies, where it has demonstrated advantages over traditional analytical methods by enabling rapid, direct analysis without disrupting the biological system (Suwa et al. 2024). This comprehensive analytical capability, spanning from subcellular to colony-level observations, has significantly enhanced our understanding of microbial metabolism and holds promise for both fundamental research and industrial applications in microbial biotechnology.

The Raman microspectroscopy-MCR analysis approach also offers a significant advancement in microbial single-cell classification through direct metabolite profiling (Ando et al. 2023). Unlike conventional machine learning approaches that rely solely on numerical spectral data, this methodology enables the identification of specific molecular markers as rational classification criteria. By integrating molecular-level information with cellular phenotypes, this approach provides a more robust and interpretable framework for understanding microbial cell states and their transitions. This chemically interpretable approach not only provides unprecedented insights into cellular states and population dynamics but also enables quantitative monitoring of temporal changes in cell populations while maintaining the advantages of nondestructive, label-free analysis at single-cell resolution.

Furthermore, the integration of Raman microspectroscopy analysis with complementary analytical techniques, particularly single-cell genomics, has opened new frontiers in microbial characterization (Kogawa et al. 2022). This multimodal approach enables simultaneous analysis of metabolomic and genomic information at single-cell resolution, offering remarkable new perspectives into complex microbial systems. By combining metabolite profiling with genomic analysis through techniques such as microdroplet-based isolation, researchers can now directly link chemical signatures to their biosynthetic origins. This integrated methodology has proven particularly valuable for investigating uncultured microorganisms and complex microbial communities, where traditional analytical approaches alone have been insufficient to establish clear structure-function relationships at the single-cell level.

6.4.4 Establishing Raman Metabolomics: From Experimental Design to Data Acquisition

Implementing Raman microspectroscopy for single-cell microbial metabolomics requires careful consideration of instrumentation, sample preparation, and data analysis (Butler et al. 2016; Lee et al. 2021a). For instrumentation, confocal Raman microspectroscopic systems typically employ monochromatic lasers (commonly 532 nm or 785 nm), high-NA objectives (100× with NA $\geq$ 1.2), and sensitive CCD detectors. The 532 nm laser provides stronger signals but may induce more fluorescence background compared to 785 nm, making the latter preferable for pigmented microbes.

Sample preparation influences spectral quality. Microbial suspensions (OD600 $\approx$ 0.5–1.0) can be mounted on glass or quartz substrates that produce minimal background signals. For environmental samples, washing with isotonic solutions helps remove interfering extracellular substances. In stable isotope probing experiments, cells are cultivated with labeled substrates (e.g., ^{13}C-glucose or D_2O) at 10–50% concentration of the normal source.

During measurements, laser powers of 1–5 mW for bacteria and exposure times of 0.5–3 seconds per spectrum help balance signal quality with minimal photodamage. Spatial mapping typically employs 0.5–1 μm steps for bacteria. Data preprocessing includes cosmic ray removal, background subtraction, normalization, and smoothing before applying multivariate analysis methods like PCA or MCR.

This streamlined workflow enables researchers to effectively implement Raman microspectroscopy for investigating metabolic heterogeneity in microbial populations at the single-cell level, though specific protocols may require optimization for particular research questions.

6.4.5 Future Perspectives

Future advances in Raman microspectroscopy and its associated analytical methods are expected to further expand their applications in microbial systems biology. The continued development of high-speed imaging capabilities, combined with improved molecular resolution and sensitivity, will enable real-time monitoring of metabolic dynamics in living cells. This evolution of the technology may provide critical insights into cellular heterogeneity and metabolic adaptation mechanisms, particularly in complex microbial communities where traditional bulk analysis methods have proven insufficient. The establishment of comprehensive spectral databases and standardized analysis protocols will be crucial for the broader adoption of these techniques, enabling reliable molecular assignments and cross-study comparisons across different laboratories and experimental conditions.

The convergence of Raman-based technologies with other analytical platforms, particularly advanced genomic and transcriptomic methods, presents exciting opportunities for comprehensive single-cell multi-omics analysis. As computational methods and data resources continue to evolve, the integration of chemically interpretable molecular profiles with other cellular parameters could revolutionize our understanding of microbial cell states and their regulation. These developments, supported by robust data infrastructure and sharing platforms, hold promise not only for fundamental microbiological research but also for practical applications in biotechnology, ranging from strain improvement for industrial fermentation to the discovery of novel bioactive compounds through rational screening of environmental microorganisms.

6.5 Conclusion

Recent advancements in single-cell omics technologies have transformed our ability to analyze microbial cells at an unprecedented resolution, providing new insights into their genetic, transcriptomic, and metabolic landscapes. These innovations have allowed researchers to explore the diversity, functionality, and ecological roles of cultured and uncultured microorganisms, contributing to a deeper understanding of microbial communities and their interactions.

One of the most significant developments in microbial single-cell genomics is the application of microfluidic droplet-based approaches. These techniques have substantially improved the efficiency and accuracy of whole-genome amplification (WGA), addressing key challenges such as amplification bias and contamination. By compartmentalizing individual cells into nanoliter or picoliter reaction chambers, these methods facilitate high-throughput sequencing of single microbial genomes, enabling large-scale analyses of complex environmental microbiomes. Furthermore, targeted single-cell genomics has enhanced the recovery of specific microbial genomes, allowing researchers to selectively analyze key bacterial and viral strains involved in host-microbe interactions and environmental processes.

Similarly, single-cell transcriptomics has provided critical insights into bacterial gene expression heterogeneity. Traditional bulk RNA sequencing has often masked cell-to-cell variability, but recent single-cell RNA sequencing (scRNA-seq) techniques have revealed dynamic gene expression patterns at the individual cell level. These findings have advanced our understanding of bacterial adaptation mechanisms, including stress responses, antibiotic-resistance, and virulence factor regulation. Although technical challenges remain—such as low RNA content, ribosomal RNA depletion, and efficient cell lysis—ongoing improvements in library preparation methods and barcoding strategies continue to expand the potential applications of bacterial scRNA-seq.

In parallel, single-cell metabolomics has emerged as a powerful tool for understanding cellular function and metabolic heterogeneity. Techniques such as Raman microspectroscopy enable label-free, high-resolution analysis of microbial metabolism, offering real-time snapshots of cellular biochemical processes. The integration of Raman spectroscopy with computational approaches like MCR analysis has allowed for precise characterization of intracellular metabolites, cellular differentiation, and secondary metabolite production in microorganisms. These advancements have been instrumental in microbial biotechnology, enabling targeted screening of metabolite-producing strains and optimizing bioproduction processes.

The integration of multi-omics approaches—combining genomics, transcriptomics, and metabolomics at the single-cell level—represents the next frontier in microbial research. By linking genomic information with functional expression data, researchers can construct comprehensive models of microbial physiology and ecosystem dynamics. The application of artificial intelligence (AI) and machine learning (ML) to multi-omics datasets is expected to further enhance data interpretation, enabling predictive modeling of microbial behavior in diverse environments.

Beyond fundamental research, single-cell omics technologies have significant implications for applied sciences, including environmental microbiology, biomedical research, and industrial biotechnology. In environmental science, these approaches help elucidate the role of microorganisms in biogeochemical cycles and ecosystem stability. In medicine, single-cell analyses provide new avenues for understanding microbiome-related diseases and developing targeted microbial therapies, such as phage therapy for antibiotic-resistant infections. Additionally, in biotechnology, these techniques facilitate the discovery of novel enzymes and bioactive compounds, accelerating advancements in synthetic biology and bioengineering.

Despite these remarkable advancements, several challenges remain. The high cost and technical complexity of single-cell omics limit their widespread adoption, particularly in resource-limited settings. Standardization of methodologies, data analysis pipelines, and validation protocols will be crucial for ensuring reproducibility and comparability across studies. Additionally, the need for improved single-cell isolation, sequencing accuracy, and high-throughput data processing remains an ongoing focus for future research.

In conclusion, single-cell omics technologies have revolutionized our understanding of microbial life, enabling the exploration of previously inaccessible biological processes. As these methodologies continue to evolve, their integration with AI-driven analytics and multi-omics frameworks will unlock new possibilities in microbial ecology, biomedicine, and biotechnology. By overcoming existing challenges and expanding their applications, single-cell omics will remain at the forefront of microbiological research, driving scientific discoveries that impact diverse fields.

References

Allen LZ, Ishoey T, Novotny MA, McLean JS, Lasken RS, Williamson SJ (2011) Single virus genomics: a new tool for virus discovery. PLoS One 6(3):e17722. https://doi.org/10.1371/journal.pone.0017722

Ando M, Hamaguchi HO (2014) Molecular component distribution imaging of living cells by multivariate curve resolution analysis of space-resolved Raman spectra. J Biomed Opt 19(1):011016. https://doi.org/10.1117/1.JBO.19.1.011016

Ando M, Sugiyama K, Kubo K, Horii S, Hano T, Tomaru Y, Takeyama H (2023) Single-cell level Raman molecular profiling reveals the classification of growth phases of Chaetoceros tenuissimus. J Phys Chem B 127(22):5027–5033. https://doi.org/10.1021/acs.jpcb.3c02152

Arnoldini M, Vizcarra IA, Pena-Miller R, Stocker N, Diard M, Vogel V, Beardmore RE, Hardt WD, Ackermann M (2014) Bistable expression of virulence genes in salmonella leads to the formation of an antibiotic-tolerant subpopulation. PLoS Biol 12(8):e1001928. https://doi.org/10.1371/journal.pbio.1001928

Bakthavatsalam S, Dodo K, Sodeoka M (2021) A decade of alkyne-tag Raman imaging (ATRI): applications in biological systems. RSC Chem Biol 2(5):1415–1429. https://doi.org/10.1039/d1cb00116g

Blattman SB, Jiang W, Oikonomou P, Tavazoie S (2020) Prokaryotic single-cell RNA sequencing by in situ combinatorial indexing. Nat Microbiol 5(10):1192–1201. https://doi.org/10.1038/s41564-020-0729-6

Brennan MA, Rosenthal AZ (2021) Single-cell RNA sequencing elucidates the structure and organization of microbial communities. Front Microbiol 12:713128. https://doi.org/10.3389/fmicb.2021.713128

Butler HJ, Ashton L, Bird B, Cinque G, Curtis K, Dorney J, Esmonde-White K, Fullwood NJ, Gardner B, Martin-Hirsch PL, Walsh MJ, McAinsh MR, Stone N, Martin FL (2016) Using Raman spectroscopy to characterize biological materials. Nat Protoc 11(4):664–687. https://doi.org/10.1038/nprot.2016.036

Chijiiwa R, Hosokawa M, Kogawa M, Nishikawa Y, Ide K, Sakanashi C, Takahashi K, Takeyama H (2020) Single-cell genomics of uncultured bacteria reveals dietary fiber responders in the mouse gut microbiota. Microbiome 8(1):5. https://doi.org/10.1186/s40168-019-0779-2

Cui D, Kong L, Wang Y, Zhu Y, Zhang C (2022) In situ identification of environmental microorganisms with Raman spectroscopy. Environ Sci Ecotechnol 11:100187. https://doi.org/10.1016/j.ese.2022.100187

Dakal TC, Dhakar R, Beura A, Moar K, Maurya PK, Sharma NK, Ranga V, Kumar A (2024) Emerging methods and techniques for cancer biomarker discovery. Pathol Res Pract 262:155567. https://doi.org/10.1016/j.prp.2024.155567

Du J, Su Y, Qian C, Yuan D, Miao K, Lee D, Ng AHC, Wijker RS, Ribas A, Levine RD, Heath JR, Wei L (2020) Raman-guided subcellular pharmaco-metabolomics for metastatic melanoma cells. Nat Commun 11(1):4830. https://doi.org/10.1038/s41467-020-18376-x

Dusny C, Lohse M, Reemtsma T, Schmid A, Lechtenfeld OJ (2019) Quantifying a biocatalytic product from a few living microbial cells using microfluidic cultivation coupled to FT-ICR-MS. Anal Chem 91(11):7012–7018. https://doi.org/10.1021/acs.analchem.9b00978

Feng X, Liu BF, Li J, Liu X (2015) Advances in coupling microfluidic chips to mass spectrometry. Mass Spectrom Rev 34(5):535–557. https://doi.org/10.1002/mas.21417

Gu W, Crawford ED, O'Donovan BD, Wilson MR, Chow ED, Retallack H, DeRisi JL (2016) Depletion of abundant sequences by hybridization (DASH): using Cas9 to remove unwanted high-abundance species in sequencing libraries and molecular counting applications. Genome Biol 17:41. https://doi.org/10.1186/s13059-016-0904-5

Hayashi T, Ozaki H, Sasagawa Y, Umeda M, Danno H, Nikaido I (2018) Single-cell full-length total RNA sequencing uncovers dynamics of recursive splicing and enhancer RNAs. Nat Commun 9(1):619. https://doi.org/10.1038/s41467-018-02866-0

Helaine S, Cheverton AM, Watson KG, Faure LM, Matthews SA, Holden DW (2014) Internalization of Salmonella by macrophages induces formation of nonreplicating persisters. Science 343(6167):204–208. https://doi.org/10.1126/science.1244705

Hockenberry AM, Micali G, Takacs G, Weng J, Hardt WD, Ackermann M (2021) Microbiota-derived metabolites inhibit *Salmonella* virulent subpopulation development by acting on single-cell behaviors. Proc Natl Acad Sci U S A 118(31):e2103027118, Article e2103027118. https://doi.org/10.1073/pnas.2103027118

Homberger C, Barquist L, Vogel J (2022) Ushering in a new era of single-cell transcriptomics in bacteria. Microlife 3:uqac020. https://doi.org/10.1093/femsml/uqac020

Homberger C, Hayward RJ, Barquist L, Vogel J (2023) Improved bacterial single-cell RNA-Seq through automated MATQ-Seq and Cas9-based removal of rRNA reads. MBio 14(2):e0355722. https://doi.org/10.1128/mbio.03557-22

Horii S, Ando M, Samuel AZ, Take A, Nakashima T, Matsumoto A, Takahashi YK, Takeyama H (2020) Detection of penicillin G produced by Penicillium chrysogenum with Raman microspectroscopy and multivariate curve resolution-alternating least-squares methods. J Nat Prod 83(11):3223–3229. https://doi.org/10.1021/acs.jnatprod.0c00214

Horii S, Samuel AZ, Nakashima T, Take A, Matsumoto A, Takahashi Y, Ando M, Takeyama H (2023) Mycelial differentiation linked avermectin production in Streptomyces avermitilis studied with Raman imaging. Appl Microbiol Biotechnol 107(1):369–378. https://doi.org/10.1007/s00253-022-12314-1

Hosokawa M, Nishikawa Y (2024) Tools for microbial single-cell genomics for obtaining uncultured microbial genomes. Biophys Rev 16(1):69–77. https://doi.org/10.1007/s12551-023-01124-y

Hosokawa M, Nishikawa Y, Kogawa M, Takeyama H (2017) Massively parallel whole genome amplification for single-cell sequencing using droplet microfluidics. Sci Rep 7(1):5199. https://doi.org/10.1038/s41598-017-05436-4

Ide K, Nishikawa Y, Maruyama T, Tsukada Y, Kogawa M, Takeda H, Ito H, Wagatsuma R, Miyaoka R, Nakano Y, Kinjo K, Ito M, Hosokawa M, Yura K, Suda S, Takeyama H (2022) Targeted single-cell genomics reveals novel host adaptation strategies of the symbiotic bacteria Endozoicomonas in Acropora tenuis coral. Microbiome 10(1):220. https://doi.org/10.1186/s40168-022-01395-9

Imdahl F, Saliba AE (2020) Advances and challenges in single-cell RNA-seq of microbial communities. Curr Opin Microbiol 57:102–110. https://doi.org/10.1016/j.mib.2020.10.001

Imdahl F, Vafadarnejad E, Homberger C, Saliba AE, Vogel J (2020) Single-cell RNA-sequencing reports growth-condition-specific global transcriptomes of individual bacteria. Nat Microbiol 5(10):1202–1206. https://doi.org/10.1038/s41564-020-0774-1

Jayan H, Pu H, Sun DW (2022) Recent developments in Raman spectral analysis of microbial single cells: techniques and applications. Crit Rev Food Sci Nutr 62(16):4294–4308. https://doi.org/10.1080/10408398.2021.1945534

Jia M, Zhu S, Xue M-Y, Chen H, Xu J, Song M, Tang Y, Liu X, Tao Y, Zhang T, Liu J-X, Wang Y, Sun H-Z (2024) Single-cell transcriptomics across 2,534 microbial species reveals functional heterogeneity in the rumen microbiome. Nat Microbiol 9(7):1884–1898. https://doi.org/10.1038/s41564-024-01723-9

Kivioja T, Vaharautio A, Karlsson K, Bonke M, Enge M, Linnarsson S, Taipale J (2011) Counting absolute numbers of molecules using unique molecular identifiers. Nat Methods 9(1):72–74. https://doi.org/10.1038/nmeth.1778

Kogawa M, Hosokawa M, Nishikawa Y, Mori K, Takeyama H (2018) Obtaining high-quality draft genomes from uncultured microbes by cleaning and co-assembly of single-cell amplified genomes. Sci Rep 8(1):2059. https://doi.org/10.1038/s41598-018-20384-3

Kogawa M, Miyaoka R, Hemmerling F, Ando M, Yura K, Ide K, Nishikawa Y, Hosokawa M, Ise Y, Cahn JKB, Takada K, Matsunaga S, Mori T, Piel J, Takeyama H (2022) Single-cell metabolite detection and genomics reveals uncultivated talented producer. PNAS Nexus 1(1):pgab007. https://doi.org/10.1093/pnasnexus/pgab007

Kuchina A, Brettner LM, Paleologu L, Roco CM, Rosenberg AB, Carignano A, Kibler R, Hirano M, DePaolo RW, Seelig G (2021) Microbial single-cell RNA sequencing by split-pool barcoding. Science 371(6531). https://doi.org/10.1126/science.aba5257

Lan F, Haliburton JR, Yuan A, Abate AR (2016) Droplet barcoding for massively parallel single-molecule deep sequencing. Nat Commun 7:11784. https://doi.org/10.1038/ncomms11784

Lan F, Demaree B, Ahmed N, Abate AR (2017) Single-cell genome sequencing at ultra-high-throughput with microfluidic droplet barcoding. Nat Biotechnol 35(7):640–646. https://doi.org/10.1038/nbt.3880

Lee KS, Landry Z, Pereira FC, Wagner M, Berry D, Huang WE, Taylor GT, Kneipp J, Popp J, Zhang M, Cheng J-X, Stocker R (2021a) Raman microspectroscopy for microbiology. Nat Rev Methods Primers 1(1). https://doi.org/10.1038/s43586-021-00075-6

Lee KS, Pereira FC, Palatinszky M, Behrendt L, Alcolombri U, Berry D, Wagner M, Stocker R (2021b) Optofluidic Raman-activated cell sorting for targeted genome retrieval or cultivation of microbial cells with specific functions. Nat Protoc 16(2):634–676. https://doi.org/10.1038/s41596-020-00427-8

Leung K, Zahn H, Leaver T, Konwar KM, Hanson NW, Page AP, Lo CC, Chain PS, Hallam SJ, Hansen CL (2012) A programmable droplet-based microfluidic device applied to multiparameter analysis of single microbes and microbial communities. Proc Natl Acad Sci U S A 109(20):7665–7670. https://doi.org/10.1073/pnas.1106752109

Leung K, Klaus A, Lin BK, Laks E, Biele J, Lai D, Bashashati A, Huang YF, Aniba R, Moksa M, Steif A, Mes-Masson AM, Hirst M, Shah SP, Aparicio S, Hansen CL (2016) Robust high-performance nanoliter-volume single-cell multiple displacement amplification on planar substrates. Proc Natl Acad Sci U S A 113(30):8484–8489. https://doi.org/10.1073/pnas.1520964113

Li W, Shao C, Li C, Zhou H, Yu L, Yang J, Wan H, He Y (2023) Metabolomics: a useful tool for ischemic stroke research. J Pharm Anal 13(9):968–983. https://doi.org/10.1016/j.jpha.2023.05.015

Liu X, Locasale JW (2017) Metabolomics: a primer. Trends Biochem Sci 42(4):274–284. https://doi.org/10.1016/j.tibs.2017.01.004

Liu R, Yang Z (2021) Single cell metabolomics using mass spectrometry: techniques and data analysis. Anal Chim Acta 1143:124–134. https://doi.org/10.1016/j.aca.2020.11.020

Ma P, Amemiya HM, He LL, Gandhi SJ, Nicol R, Bhattacharyya RP, Smillie CS, Hung DT (2023) Bacterial droplet-based single-cell RNA-seq reveals antibiotic-associated heterogeneous cellular states. Cell 186(4):877–891.e814. https://doi.org/10.1016/j.cell.2023.01.002

Marcy Y, Ishoey T, Lasken RS, Stockwell TB, Walenz BP, Halpern AL, Beeson KY, Goldberg SM, Quake SR (2007) Nanoliter reactors improve multiple displacement amplification of genomes from single cells. PLoS Genet 3(9):1702–1708. https://doi.org/10.1371/journal.pgen.0030155

Martinez Martinez J, Martinez-Hernandez F, Martinez-Garcia M (2020) Single-virus genomics and beyond. Nat Rev Microbiol 18(12):705–716. https://doi.org/10.1038/s41579-020-00444-0

Martinez-Hernandez F, Fornas O, Lluesma Gomez M, Bolduc B, de la Cruz Pena MJ, Martinez JM, Anton J, Gasol JM, Rosselli R, Rodriguez-Valera F, Sullivan MB, Acinas SG, Martinez-Garcia M (2017) Single-virus genomics reveals hidden cosmopolitan and abundant viruses. Nat Commun 8:15892. https://doi.org/10.1038/ncomms15892

McNulty R, Sritharan D, Pahng SH, Meisch JP, Liu S, Brennan MA, Saxer G, Hormoz S, Rosenthal AZ (2023) Probe-based bacterial single-cell RNA sequencing predicts toxin regulation. Nat Microbiol 8(5):934–945. https://doi.org/10.1038/s41564-023-01348-4

Nicholson JK, Lindon JC, Holmes E (1999) 'Metabonomics': understanding the metabolic responses of living systems to pathophysiological stimuli via multivariate statistical analysis of biological NMR spectroscopic data. Xenobiotica 29(11):1181–1189. https://doi.org/10.1080/004982599238047

Nishikawa Y, Hosokawa M, Maruyama T, Yamagishi K, Mori T, Takeyama H (2015) Monodisperse picoliter droplets for low-bias and contamination-free reactions in single-cell whole genome amplification. PLoS One 10(9):e0138733. https://doi.org/10.1371/journal.pone.0138733

Nishikawa Y, Kogawa M, Hosokawa M, Wagatsuma R, Mineta K, Takahashi K, Ide K, Yura K, Behzad H, Gojobori T, Takeyama H (2022) Validation of the application of gel beads-based single-cell genome sequencing platform to soil and seawater. ISME Commun 2(1):92. https://doi.org/10.1038/s43705-022-00179-4

Nishikawa Y, Wagatsuma R, Tsukada Y, Chia-Ling L, Chijiiwa R, Hosokawa M, Takeyama H (2024) Large-scale single-virus genomics uncovers hidden diversity of river water viruses and diversified gene profiles. ISME J 18(1). https://doi.org/10.1093/ismejo/wrae124

Nishimura M, Takeyama H, Hosokawa M (2023) Enhancing the sensitivity of bacterial single-cell RNA sequencing using RamDA-seq and Cas9-based rRNA depletion. J Biosci Bioeng 136(2):152–158. https://doi.org/10.1016/j.jbiosc.2023.05.010

Oyola SO, Manske M, Campino S, Claessens A, Hamilton WL, Kekre M, Drury E, Mead D, Gu Y, Miles A, MacInnis B, Newbold C, Berriman M, Kwiatkowski DP (2014) Optimized whole-genome amplification strategy for extremely AT-biased template. DNA Res 21(6):661–671. https://doi.org/10.1093/dnares/dsu028

Pezzotti G (2021) Raman spectroscopy in cell biology and microbiology. J Raman Spectrosc 52(12):2348–2443. https://doi.org/10.1002/jrs.6204

Prezza G, Heckel T, Dietrich S, Homberger C, Westermann AJ, Vogel J (2020) Improved bacterial RNA-seq by Cas9-based depletion of ribosomal RNA reads. RNA 26(8):1069–1078. https://doi.org/10.1261/rna.075945.120

Pryszlak A, Wenzel T, Seitz KW, Hildebrand F, Kartal E, Cosenza MR, Benes V, Bork P, Merten CA (2022) Enrichment of gut microbiome strains for cultivation-free genome sequencing using droplet microfluidics. Cell Rep Methods 2(1):None. https://doi.org/10.1016/j.crmeth.2021.100137

Puppels GJ, de Mul FF, Otto C, Greve J, Robert-Nicoud M, Arndt-Jovin DJ, Jovin TM (1990) Studying single living cells and chromosomes by confocal Raman microspectroscopy. Nature 347(6290):301–303. https://doi.org/10.1038/347301a0

Qin S, Miao D, Zhang X, Zhang Y, Bai Y (2023) Methods developments of mass spectrometry based single cell metabolomics. TrAC Trends Anal Chem 164. https://doi.org/10.1016/j.trac.2023.117086

Rhee M, Light YK, Meagher RJ, Singh AK (2016) Digital droplet multiple displacement amplification (ddMDA) for whole genome sequencing of limited DNA samples. PLoS One 11(5):e0153699. https://doi.org/10.1371/journal.pone.0153699

Sampaio NMV, Blassick CM, Andreani V, Lugagne JB, Dunlop MJ (2022) Dynamic gene expression and growth underlie cell-to-cell heterogeneity in Escherichia coli stress response. Proc Natl Acad Sci U S A 119(14):e2115032119. https://doi.org/10.1073/pnas.2115032119

Samuel AZ, Horii S, Nakashima T, Shibata N, Ando M, Takeyama H (2022) Raman microspectroscopy imaging analysis of extracellular vesicles biogenesis by filamentous fungus Penicilium chrysogenum. Adv Biol (Weinh) 6(6):e2101322. https://doi.org/10.1002/adbi.202101322

Schirmer M, Dusny C (2023) Microbial single-cell mass spectrometry: status, challenges, and prospects. Curr Opin Biotechnol 83:102977. https://doi.org/10.1016/j.copbio.2023.102977

Schirmer M, Wink K, Ohla S, Belder D, Schmid A, Dusny C (2020) Conversion efficiencies of a few living microbial cells detected at a high throughput by droplet-based ESI-MS. Anal Chem 92(15):10700–10708. https://doi.org/10.1021/acs.analchem.0c01839

Shen Y, Qian Q, Ding L, Qu W, Zhang T, Song M, Huang Y, Wang M, Xu Z, Chen J, Dong L, Chen H, Shen E, Zheng S, Chen Y, Liu J, Fan L, Wang Y (2024) High-throughput single-microbe RNA sequencing reveals adaptive state heterogeneity and host-phage activity associations in human gut microbiome. Protein Cell. https://doi.org/10.1093/procel/pwae027

Suwa S, Ando M, Nakashima T, Horii S, Anai T, Takeyama H (2024) In situ Raman hyperspectral analysis of microbial colonies for secondary metabolites screening. Anal Chem 96(37):14909–14917. https://doi.org/10.1021/acs.analchem.4c02906

Tang F, Barbacioru C, Wang Y, Nordman E, Lee C, Xu N, Wang X, Bodeau J, Tuch BB, Siddiqui A, Lao K, Surani MA (2009) mRNA-Seq whole-transcriptome analysis of a single cell. Nat Methods 6(5):377–382. https://doi.org/10.1038/nmeth.1315

Tang X, Wu Q, Shang L, Liu K, Ge Y, Liang P, Li B (2024) Raman cell sorting for single-cell research. Front Bioeng Biotechnol 12:1389143. https://doi.org/10.3389/fbioe.2024.1389143

Tauler R (1995) Multivariate curve resolution applied to second order data. Chemom Intel Lab Syst 30(1):133–146. https://doi.org/10.1016/0169-7439(95)00047-x

Taylor MJ, Lukowski JK, Anderton CR (2021) Spatially resolved mass spectrometry at the single cell: recent innovations in proteomics and metabolomics. J Am Soc Mass Spectrom 32(4):872–894. https://doi.org/10.1021/jasms.0c00439

Wagatsuma R, Nishikawa Y, Hosokawa M, Takeyama H (2025) vClean: assessing virus sequence contamination in viral genomes. NAR Genom Bioinform 7(1):lqae185. https://doi.org/10.1093/nargab/lqae185

Wang Y, Huang WE, Cui L, Wagner M (2016) Single cell stable isotope probing in microbiology using Raman microspectroscopy. Curr Opin Biotechnol 41:34–42. https://doi.org/10.1016/j.copbio.2016.04.018

Wang B, Lin AE, Yuan J, Novak KE, Koch MD, Wingreen NS, Adamson B, Gitai Z (2023a) Single-cell massively-parallel multiplexed microbial sequencing (M3-seq) identifies rare bacterial populations and profiles phage infection. Nat Microbiol 8(10):1846–1862. https://doi.org/10.1038/s41564-023-01462-3

Wang Z, Zhu H, Xiong W (2023b) Advances in mass spectrometry-based multi-scale metabolomic methodologies and their applications in biological and clinical investigations. Sci Bull (Beijing) 68(19):2268–2284. https://doi.org/10.1016/j.scib.2023.08.047

Wang Z, Li S, Zhang S, Zhang T, Wu Y, Liu A, Wang K, Ji X, Cao H, Zhang Y, Tan EK, Wang Y, Wang Y, Liu W (2024) Hosts manipulate lifestyle switch and pathogenicity heterogeneity of opportunistic pathogens in the single-cell resolution. Elife 13. https://doi.org/10.7554/eLife.96789

Xu L, Brito IL, Alm EJ, Blainey PC (2016) Virtual microfluidics for digital quantification and single-cell sequencing. Nat Methods 13(9):759–762. https://doi.org/10.1038/nmeth.3955

Xu J, Yu T, Zois CE, Cheng JX, Tang Y, Harris AL, Huang WE (2021) Unveiling cancer metabolism through spontaneous and coherent Raman spectroscopy and stable isotope probing. Cancer 13(7). https://doi.org/10.3390/cancers13071718

Xu Z, Wang Y, Sheng K, Rosenthal R, Liu N, Hua X, Zhang T, Chen J, Song M, Lv Y, Zhang S, Huang Y, Wang Z, Cao T, Shen Y, Jiang Y, Yu Y, Chen Y, Guo G et al (2023) Droplet-based high-throughput single microbe RNA sequencing by smRandom-seq. Nat Commun 14(1):5130. https://doi.org/10.1038/s41467-023-40137-9

Yue S, Cheng JX (2016) Deciphering single cell metabolism by coherent Raman scattering microscopy. Curr Opin Chem Biol 33:46–57. https://doi.org/10.1016/j.cbpa.2016.05.016

Zhang J, Su X, Wang Y, Wang X, Zhou S, Jia H, Jing X, Gong Y, Wang J, Xu J (2023) Improved single-cell genome amplification by a high-efficiency phi29 DNA polymerase. Front Bioeng Biotechnol 11:1233856. https://doi.org/10.3389/fbioe.2023.1233856

Zheng W, Zhao S, Yin Y, Zhang H, Needham DM, Evans ED, Dai CL, Lu PJ, Alm EJ, Weitz DA (2022) High-throughput, single-microbe genomics with strain resolution, applied to a human gut microbiome. Science 376(6597):eabm1483. https://doi.org/10.1126/science.abm1483

Zucha D, Androvic P, Kubista M, Valihrach L (2020) Performance comparison of reverse transcriptases for single-cell studies. Clin Chem 66(1):217–228. https://doi.org/10.1373/clinchem.2019.307835

Challenges and Future Perspectives for Single-Cell Multi-Omics in Prokaryotes

7

Tyra Magold and Anne-Kristin Kaster

Abstract

Omics technologies, aka genomics, transcriptomics, proteomics, and metabolomics are powerful tools to elucidate microbial and functional dark matter. Single-cell technologies have aided in the understanding of heterogeneity and phenotypic differences of and between cells in microbial communities, enabling new insights into diversity, evolutionary histories, metabolic pathways, and potentials for biotechnological applications. Nevertheless, to fully understand single-cell biology in a complete and comprehensive way, multiple 'omics technologies have to be combined and applied simultaneously to one individual cell. Single-cell multi-omics methods have been used in eukaryotes for the last ten years; however, they still face challenges and limitations for applications in prokaryotes. In this chapter, we describe state-of-the-art technologies and future perspectives for the use of single-cell multi-omics in bacteria and archaea.

Keywords

Microbial Dark Matter · Functional Dark Matter · Genomics · Transcriptomics · Proteomics · Metabolomics

---

T. Magold · A.-K. Kaster (✉)
Institute for Biological Interfaces 5 (IBG-5), Karlsruhe Institute of Technology, Eggenstein-Leopoldshafen, Germany
e-mail: kaster@kit.edu

C. Rinke, M. Džunková (eds.), *Single-Cell Omics in Microbiome Research*, The Microbiomes of Humans, Animals, Plants, and the Environment 7,
https://doi.org/10.1007/978-3-032-07527-7_7

7.1 Single-Cell Technologies: Status Quo, Challenges, and Limitations

Prokaryotes are highly diverse and abundant organisms with significant potential for medical or biotechnological applications. However, an estimated 99% of all microbial species are currently uncultured, the so-called microbial dark matter (MDM) (McDonald et al. 2012; Lloyd et al. 2018; Kaster and Sobol 2020). Cultivation-independent "omics" methods have therefore become the gold standard for studying MDM (Lok 2015; Jiao et al. 2021; Vollmers et al. 2022, 2023). Metagenomics is the extraction and shotgun sequencing of DNA, followed by bioinformatic assembly and binning approaches to gain information about phylogenetic diversity, abundance, and genetic composition of a sample (Hugenholtz and Tyson 2008; Vollmers et al. 2017, 2023). Correspondingly, metatranscriptomics is the extraction and sequencing of RNA to then link expression profiles of genes to certain environmental conditions or stimuli (Carvalhais et al. 2012; Ojala et al. 2023). Metaproteomics investigates proteins (e.g., their functions and interactions) by using mass spectrometry (MS) for amino acid analysis (Grassl et al. 2016; Van Den Bossche et al. 2021; Ahmad and Budnik 2023). Metabolomics workflows can include fluorescence or isotope labeling, followed by either fluorescence intensity measurements, vibrational spectroscopy (e.g., Raman) or MS (Mozzi et al. 2013; Premasiri et al. 2016; Zhan et al. 2023). All four "omics" methods have massively improved our understanding of microorganisms regarding their diversity, evolutionary histories, metabolic pathways, and interactions with nature. Nevertheless, meta-omics are often limited by their inability to provide high-resolution, cell-specific insights, and certain traits can often not be tracked back to the corresponding microorganisms in bulk samples. Hence, the diversity, but also heterogeneity of different organisms and their responses to environmental changes or interactions with each other are often masked. This holds especially true for microorganisms with low abundance in a habitat and/or samples with high strain diversity (Podar et al. 2007; Wang and Navin 2015; Dam et al. 2020).

These knowledge gaps can be addressed by using single-cell "omics," complementing meta-omics technologies (Wang and Navin 2015; Robinson et al. 2021). Unfortunately, methods for prokaryotes often lag behind the ones for eukaryotes, especially mammalian cells, for various reasons. In order to better understand the cause for specific limitations and challenges when single-cell omics technologies are applied to bacteria and archaea, differences in cell types regarding cell membrane, cell size and volume, as well as DNA, RNA, protein, and metabolite content compared to mammalian cells are listed in Table 7.1, since the vast majority of single-cell "omics" methods have been developed for human cell lines.

Established single-cell experiments usually start with sample collection and preparation (Fig. 7.1). Preparing populations for single-cell studies differs depending on the downstream applications as well as the growth forms of the prokaryotes, e.g., biofilms or floc-forming populations require additional enzymatic or mechanical dissociation of viable cells (Lee et al. 2020). For high-throughput single-cell separation, fluorescence-activated cell sorting (FACS) is most commonly used

Table 7.1 Comparison of mammalian and prokaryotic cell types. Differences in the cell type are often the cause for specific limitations and challenges in the respective workflows for prokaryotic single-cell omics

	Mammalian cell	Prokaryotic cell	Reference
Membrane	Phospholipid bilayer	Phospholipid bilayer Highly diverse variants such as peptidoglycan (10% teichoic acids) or pseudomurein in cell walls In archaea, S-layer, aka single (glyco-) protein layer Sometimes glycocalyx and/or capsules Sometimes spore-forming	Cooper (2000), Silhavy et al. (2010)
Cell size	~ 10–30 μm diameter	~ 0.5–5 μm length	Milo and Phillips (2015)
Cell volume	~ 1 pL	~ 1 fL	Milo and Phillips (2015)
DNA	~ 5–10 pg ~ 1–10 Gbp (diploid genome, excl. mitochondrial DNA) Located in the nucleus	~ 1–10 fg ~ 0.5–14 Mbp chromosome Freely floating in the cell Multiple plasmids possible	Gregory (2005), Milo and Phillips (2015), Brennan and Rosenthal (2021)
RNA	~ 10–30 pg	~ 10–100 fg	Kang et al. (2011), Milo and Phillips (2015)
mRNA	~ 3–5% of total RNA Half-life up to >24 h (median ~ 10 hours in *H. sapiens*) Polyadenylated Monocistronic	~ 2% of total RNA Half-life <10 min Non-polyadenylated Polycistronic	Tourrière et al. (2002), Brennan and Rosenthal (2021), Alberts et al. (2022)
Proteins	~ 10^{13}	~ 3×10^{6}	Milo and Phillips (2015)
Metabolites	~ 10^{-8}–1 M per metabolite ~ 100–1500 different metabolites per cell	~ 10^{-4}–10 mM per metabolite ~ 100–200 different metabolites in *E. coli* ~ 250 mM per cell	Milo and Phillips (2015), Wishart et al. (2022), Dekkers et al. (2022)

(Gross et al. 2015). This method also allows for additional target labeling prior to the standard sorting process, e.g., by fluorescence in situ hybridization (FISH) (Pratscher et al. 2018; Dam et al. 2020), or via other techniques (Hatzenpichler et al. 2020; Lo et al. 2023). These labeling approaches enable researchers to selectively sort specific taxa or organisms with a certain function.

Sorting cells via FACS involves high pressures, creating shear forces and putting stress on the cells. Some cells may therefore lyse prematurely, bearing the risk of cross-contamination and decreasing overall efficiency. In addition, this stress might cause changes in transcriptomic or metabolomic profiles. Separation to the

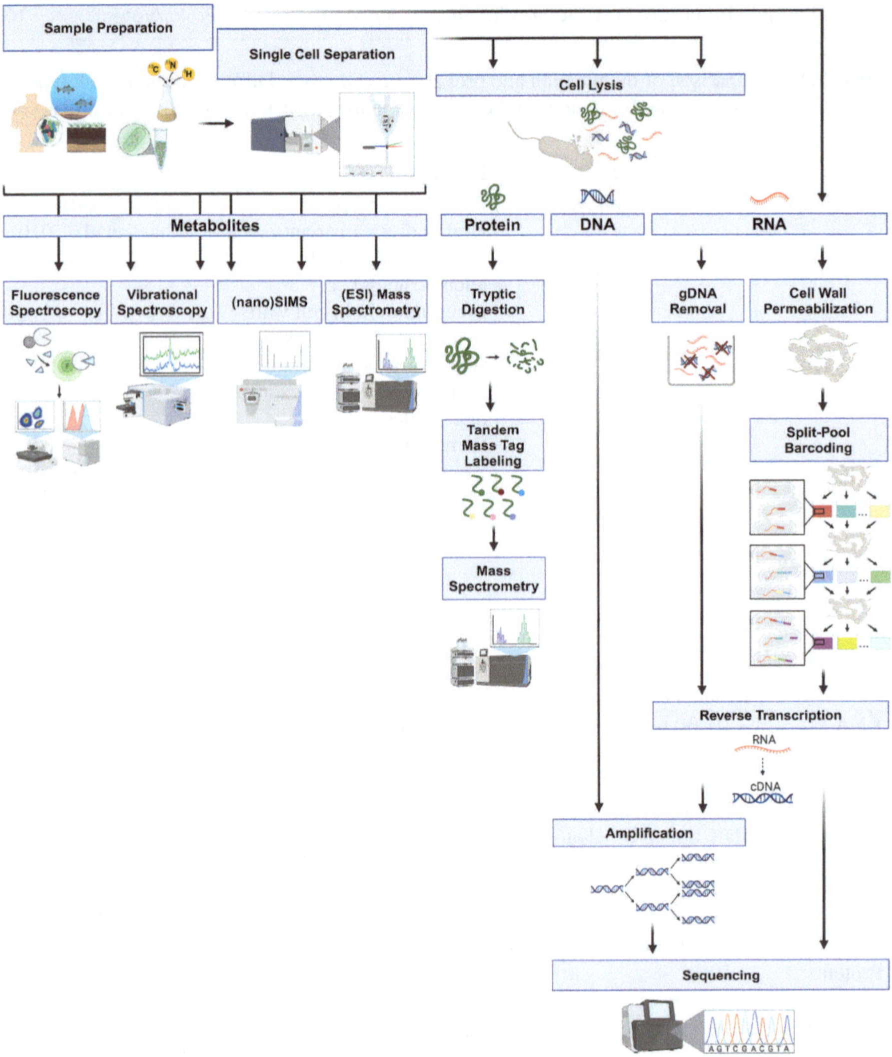

Fig. 7.1 Workflows for microbial single-cell omics. Sample preparation can include filtration, washing, labeling, or other treatments, depending on the sample source and downstream applications. Additionally, target phylogenetic or metabolic labeling of samples can be conducted. Individual single-cells are usually separated and sorted by fluorescence-activated cell sorting (FACS). Single-cell proteomics (SCP), single-cell genomics (SCG), and single-cell transcriptomics (SCT) workflows usually include a cell lysis step after sorting. SCP continues with tryptic digestion and tandem mass tag (TMT) labeling before final analysis via mass spectrometry (MS) takes place. The SCG workflow continues with a whole-genome amplification (e.g., via multiple displacement amplification), library preparation and sequencing. The SCT workflow includes gDNA removal and reverse transcription before amplification, and again, library preparation and sequencing. Another approach includes cell wall permeabilization of multiple cells, which are then subjected to split-pool barcoding, followed by reverse transcription, library preparation, and sequencing. Single cells are then bioinformatically separated based on their unique barcodes. Single-cell metabolomics (SCM) can be conducted through specific targeting of metabolites and measuring of fluorescence intensity, or via stable isotope labeling-based vibrational spectroscopy (e.g., Raman), secondary ion mass spectrometry (SIMS), nanoSIMS, or electrospray ionization mass spectrometry (ESI-MS). (Adapted after Münch et al. (2023) and Sturm et al. (2023). Figure created with Biorender.com)

individual cell level can also be achieved by cell printing, which does not require high pressures (Gross et al. 2013; Riba et al. 2016; Wiegand et al. 2021) or optical tweezers (Marshall et al. 2012; Blainey 2013), however, with much lower throughput. To analyze DNA, RNA, and protein content, the cell needs to be lysed after sorting (Fig. 7.1). This is a critical step, often struggling with the diversity of microbial cell wall types in complex samples (Table 7.1) (Rinke et al. 2014; Kaster and Sobol 2020). Applying one specific lysis method to a highly diverse sample usually leads to biased results (Islam et al. 2017; Chen et al. 2020). Sometimes, nonmechanical and mechanical methods can be combined to increase the efficiency of lysis (Anand et al. 2007). Furthermore, downstream analyses must also be taken into account when choosing a suitable lysis method, e.g., denaturation should be avoided when studying enzymes. Due to the short half-life time of just mere minutes for prokaryotic mRNA (Table 7.1), long incubation times or extra clean-up steps might also not be suitable. Further challenges emerge from the generally low ranges of DNA, RNA, proteins, and metabolites present in prokaryotic cells (Table 7.1), increasing the susceptibility of experiments to contamination. In addition, the detection limit of standard instruments is often reached at these low molecular levels.

Nevertheless, the field of microbial single-cell genomics (SCG) has vastly advanced over the past years. A key step was the invention of the so-called whole-genome amplification (Zhang et al. 1992; Raghunathan et al. 2005; Blainey 2013; Rinke et al. 2014), which allowed the minute amounts of DNA (but also RNA for whole transcriptome amplification) in the femtogram range to be amplified to nano- or even micrograms by a specific polymerase and subsequently sequenced. Unfortunately, these amplifications are costly and prone to bias. The varying GC content in different organisms causes different amplification affinities and hence efficiency (Stepanauskas and Sieracki 2007; Sabina and Leamon 2015). Another factor—also complicating downstream analysis—is the so-called chimera formation. Here, the polymerase terminates at a highly conserved region, and the elongation is continued at another genome location. This is particularly an issue for 16S rRNA, plasmids, as well as repeats (Lasken and Stockwell 2007; Acuña-Amador et al. 2018). This biased amplification reaction is the reason why SCG still suffers from uneven genome coverage, which until recently, was often below 30% with standard amplification methods. Sobol et al. published a protocol where the reaction volume was significantly reduced, leading to less contamination, cost reduction, as well as higher genome coverage up to ~90% in *Escherichia coli* cells (Sobol et al. 2024). Bowers et al. could show that a method called scMicrobe PTA achieved around 81% median genome completeness (Bowers et al. 2024). The progress in method development in SCG over the past 10 years has enabled researchers to add reference genomes of more than 40 uncultivated phylum-level clades and uncharacterized branches in the tree of life, emphasizing the importance of this method for today's microbiological research (Rinke et al. 2013; Lasken and McLean 2014; Woyke et al. 2017; Kaster and Sobol 2020; Wiegand et al. 2023).

The methods for single-cell transcriptomics (SCT) for prokaryotes have been lagging behind SCG (Münch et al. 2023). Kang et al. published the first partial transcriptome analysis of a microbial single-cell in 2011, performed via microarray (Kang et al. 2011). Four years later, Wang et al. achieved SCT of six individual *Synechocystis* sp. cells by actual RNA-sequencing (Wang et al. 2015). Recently, novel methods using microfluidics and in situ barcoding have massively improved throughput (Fig. 7.1). Here, millions of cells from a sample get tagged with specific barcodes by combinatorial indexing multiple times, resulting in unique barcode combinations for each cell based on simple probability (Blattman et al. 2020; Kuchina et al. 2021; Gaisser et al. 2024). This allows cells to be sequenced in a pool instead of relying on single cells, circumventing the need for whole transcriptome amplification. The different cells can then later be bioinformatically separated, based on their unique barcodes. Unfortunately, this method is still suffering from substantial cell loss during preparation and processing (~75%), as well as cross-contamination (Blattman et al. 2020; Wang et al. 2023). Furthermore, this method cannot be used on MDM, since transcripts have to be mapped back to known genome sequences (Haque et al. 2017; Kuchina et al. 2021). In addition, the minute amounts and low stability of mRNA in a cell persist as a challenge (Table 7.1). Approaches for eukaryotes often solve this issue by selectively enriching for mRNA through the 3′ end poly-adenosine stretch (Wang and Navin 2015; Vandereyken et al. 2023), which is not possible in microbial SCT, since prokaryotic mRNA lacks the poly-A tail. The total RNA content of a bacterium can furthermore consist of up to 98% rRNA (Rosenow et al. 2001), often resulting in rRNA reads in prokaryotic transcriptomics data. CRISPR/Cas9-mediated rRNA removal can decrease those rRNA reads by up to ~30% (Homberger et al. 2023). This method is, however, not applicable to uncharacterized species. The prokaryotic-specific challenge of polycistronic messenger RNA (Table 7.1) further complicate the downstream analysis of sequencing data, since accurate transcript lengths and quantities are crucial for correctly merging mRNA information with protein analyses and environment-specific expression patterns (Westermann and Vogel 2021).

Protein levels and composition in single cells are very dynamic and change in response to environmental conditions. In prokaryotes, they are many orders of magnitude lower than in mammalian cells. The same holds true for metabolites, which reflect a cell's physiological state, being the key constituents of intracellular reactions (Table 7.1) (Milo and Phillips 2015; Guo et al. 2021). The detection and quantification of these molecules in single cells is, however, key for a detailed understanding of an organism's metabolism in the environment, including interactions, binding and localization of molecules (Liu et al. 2016). The nature of single-cell proteomics (SCP) and metabolomics (SCM) generally requires a different workflow than SCG and SCT (Fig. 7.1), since these molecules cannot be amplified. This means that an experiment must lead to successful results with only the amounts of molecules present in the cell at the current state, posing a significant challenge for method development. Numerous advances have been made in the field of SCP for eukaryotic cells in the past years. A revolutionary step in this field was the development of high-throughput Single Cell ProtEomics by Mass Spectrometry (SCoPE-MS) in 2018 (Budnik et al. 2018). MS is a highly sensitive method, where the reporter ions result in different weights and are thus distinguishable (Tajik et al.

2022). However, the detection limits for MS are extremely dependent on the type of analyte (pure solvent or real-life samples), as well as the type of spectrometer. Detection limits in the fg/mL range are possible, however, only when using a very clean compound in a pure solvent (Thomson 2012; Beattie and Jones 2023). Recently, Végvári et al. conducted the first study on MS-based proteomics of bacterial single-cells (Végvári et al. 2023). Here, *E. coli* cells were sorted, and trypsin was added to each single cell, followed by labeling through different isobaric tandem mass tags (TMT). The samples were then multiplexed and subjected to LC-MS/MS analysis (Fig. 7.1), which combines the physical separation technology of high-performance liquid chromatography (HPLC) and the capabilities of MS. Peptides and proteins were then identified by searching against a public database. 29 peptides from 19 different proteins could be identified across the different single cells, and quantitative differentiation between one and two cells was possible, showing that SCP for known prokaryotes might be feasible in the future.

SCM for prokaryotes can be conducted with different technologies (Fig. 7.1). A targeted method is the measurement of metabolites after labeling with fluorescent probes, such as labeled nucleic acids or small molecules. Here, target metabolite concentrations are correlated with fluorescence intensity through flow cytometry or microscopy (Amantonico et al. 2010; Hong et al. 2021). Nucleic acid probes can be used to identify mRNAs, e.g., through FISH (Amann and Fuchs 2008), or change conformation when encountering a target metabolite and then turn on fluorescence (Paige et al. 2012). Small-molecule probes are fluorescent analogs of molecules like glucose, which are used to detect the uptake capacity of a cell (Natarajan and Srienc 1999). Since incubation is required before measurements, this method cannot be applied to unknown molecules or MDM. Another method to analyze metabolites in single-cells is vibrational spectroscopy (Huang et al. 2004; Zhan et al. 2023), e.g., Raman spectroscopy was used in combination with stable isotope labeling to track the uptake of specific compounds (Wang et al. 2016; Xu et al. 2017). However, multiple studies emphasize the difficulties in establishing a link between specific metabolites and the corresponding Raman spectra (Kumar et al. 2016; Cui et al. 2018; Zhan et al. 2023). Another method is MS, also in combination with stable isotope labeling. Unfortunately, the same limitations apply to the detection range of the instrument as for SCP (Schmid et al. 2010; Milo and Phillips 2015; Dusny 2022). The development of secondary ion mass spectrometry (SIMS) and nano-SIMS increased sensitivity to measure metabolites in a single bacterial cell (Bonnin and Rizzoli 2020), also for MDM, however, direct metabolite identification gets hindered by the generation of molecular fragments due to the high-energy ion beams (Zenobi 2013; Nikolic et al. 2017). The recent application of electrospray ionization mass spectrometry (ESI-MS) for prokaryotic SCM is another approach, which could show accurate molecular identification, high specificity, and high metabolite coverage (Zhang et al. 2023). Yet until now, application was only possible with a special cell form (filamentous *E. coli*), which is one order of magnitude larger than typical bacterial cells (Milo and Phillips 2015). Additionally, ESI-MS is a destructive method with low throughput, making other downstream experiments impossible. An untargeted, indirect method for SCM is through mRNA sequencing, where the changes of metabolite-related mRNA are analyzed and conclusions are drawn based on those results (Imdahl et al. 2020; Zhan et al. 2023). However, due to the fast degradation rate of prokaryotic mRNA and its polycistronic structure, this analysis does not portray a full picture of the metabolites in a cell (not shown in Fig. 7.1).

7.2 Single-Cell Multi-Omics in Prokaryotes

While single-cell mono-omics techniques (genomics, transcriptomics, proteomics, and metabolomics) have greatly enhanced our understanding of microorganisms, these methods exclude the interplay between different molecular layers and fall short of providing a comprehensive understanding of cellular function and behavior. Single-cell multi-omics approaches (connecting multiple mono-omics simultaneously on one individual cell) are therefore highly needed. This becomes especially important when investigating the so-called functional dark matter (FDM). FDM describes protein-encoding genes with unknown molecular function (Escudeiro et al. 2022; Ardern et al. 2023), also referred to as hypothetical proteins (HPs) (Galperin and Koonin 2004). Even in one of the world's best-studied model organisms, *Escherichia coli* K-12, the genome consists of over 30% of protein-encoding genes without experimental validation (Ghatak et al. 2019; Vincent 2024) and over 2% do not have sequence homologies to known proteins. In MDM, genes with unknown functions are even more abundant, with up to 60% HP-encoding genes in bacterial genomes and up to 80% in archaeal genomes (Escudeiro et al. 2022). Combining the layers of genomes, transcripts, proteins, and metabolites of cells into a joint workflow could lead to the discovery of novel features and functions to elucidate this FDM (Ardern et al. 2023). Here, several technical problems need to be solved, e.g., SCT relies on mapping the sequencing results back to the respective reference genome (Haque et al. 2017; Kuchina et al. 2021). The lack of genomic information from MDM, however, prevents mapping, underscoring the necessity of capturing DNA alongside the other molecular layers of single-cells.

Another aspect of capturing a cell's complete nature is understanding so-called posttranslational modifications (PTMs). PTM describes the modification of proteins after translation and is a common mechanism for the regulation of metabolic activity (Baysoy et al. 2023). This allows the cell to react and adapt to changes of, e.g., environmental parameters. Capturing the transcriptome and proteome of a cell simultaneously would allow for connecting different expression profiles and PTM patterns to better manipulate cells, e.g., biotechnological applications (Macek et al. 2019). By analyzing the genomic data of a single cell alongside its metabolites, novel microorganisms with potential for discoveries in pharmaceutical research could be discovered. Correlating taxonomic information with metadata would further enable the discovery of so-called "silent biosynthetic gene clusters (BGCs)", referring to genes encoding for natural products that are not expressed under laboratory conditions, since stress factors and interactions with other microorganisms of an environmental community are missing. These BGCs hold the promise for novel drug developments, e.g., antibiotics (Rutledge and Challis 2015). Furthermore, the approach of multi-omics and metadata correlation could also aid in the cultivation of MDM by obtaining detailed insights into an organism's requirements for survival and growth, enabling previously inaccessible microorganisms to be studied in laboratory settings.

7.2.1 Genomics & Transcriptomics

Combining SCG and SCT has already been established in the field of eukaryotic research in 2014 (Picelli et al. 2014; Han et al. 2014). However, no single-cell dual method of SCG and SCT has been developed for prokaryotes so far. For such methods, the separation of DNA and RNA is crucial for analysis, since the RNA needs to be reverse transcribed into cDNA. In eukaryotic single-cell workflows, five different principles exist to tackle the problem of discriminating DNA from RNA. The first option is targeted capturing of RNA via the poly-adenylated tails (polyA tail) of mRNA using oligo(dT)-coated beads. The other option is a two-step lysis of the cell, where the cell membrane, but not the nuclear membrane, is lysed. The nucleus is then separated from the cytoplasm and lysed as well (Zachariadis et al. 2020). This generates two libraries, one for DNA (nuclear membrane lysis) and one for RNA (cell membrane lysis). The third joint SCG and SCT approach follows the pre-amplification and split principle: After complete cell lysis, the mRNA gets reverse transcribed using poly-T primer, followed by joint amplification of the cDNA and gDNA. The pre-amplification reaction is then split into two, resulting in two tubes containing cDNA and gDNA. The resulting tubes are treated differently based on the target (DNA or RNA), again making use of the polyA tail (Dey et al. 2015). Due to the properties of prokaryotic cells and mRNA (Table 7.1), those approaches are not feasible for bacteria and archaea. The fourth principle is combinatorial indexing, where millions of cells get distributed into different samples and tagged with specific barcodes. This "split-barcoding" is repeated until each cell statistically obtains a unique barcode consisting of multiple sample-specific barcodes. This method uses a transposase for the first round of DNA barcoding, and tags mRNA using barcode and UMI-bearing polyT primers during RT (Yin et al. 2019). This combinatorial indexing is already used in single-cell transcriptomics in prokaryotes (Blattman et al. 2020; Kuchina et al. 2021), however, not suitable for MDM (see above) (Fig. 7.1). The fifth principle is termed "seq-split". After cell lysis, DNA and mRNA separation happen through differential barcoding for each nucleic acid species. This method also makes use of a transposase and custom adaptors, including a "DNA-barcode" and a unique molecular identifier (UMI) to fragment and tag the DNA. The whole RNA is then barcoded during reverse transcription (RT) using RT primers adapted from the so-called MATQ-seq protocol (Sheng et al. 2017). Here, the primers bind to the internal regions of all RNA transcripts and include an "RNA-barcode" and a UMI. The barcodes and UMIs can then be computationally separated by analyzing the resulting reads from the joint library preparation and sequencing (Yu et al. 2023). This method could, in theory, also be applied to prokaryotes; however, incubation times due to the shorter half-life of mRNA would probably need to be adapted. An improved version of the MATQ-seq method includes a more efficient reverse transcriptase and incorporates a CRISPR-Cas9-based rRNA depletion step. It has already been used in single-cell transcriptomics in prokaryotes (Homberger et al. 2023). Cas9-based rRNA depletion, however, requires guide RNAs and hence prior knowledge of the organism's genome, and is therefore again not suitable for MDM.

In general, if gDNA and cDNA from a single cell are present in the same vial and are subjected to amplification, decoding which strand originated from RNA is only possible with prior tagging of either RNA or DNA, to later decipher them computationally (Vandereyken et al. 2023). A method to indicate which amplified strand originated from RNA or DNA could, e.g., be done by barcoding through primers (see above), or by adding labeled nucleotides during reverse transcription (Schena et al. 1995). However, there would have to be a way for the polymerase to distinguish between labelled and unlabeled nucleotides. The incorporation of labeled nucleotides into DNA and their further amplification by polymerase chain reaction (PCR) has been previously shown (Anderson et al. 2005). Here, the research focus was mostly on the unbiased incorporation and amplification of labeled nucleotides. A polymerase capable of differentiating between labeled and unlabeled nucleotides and amplifying either one of the two would be required for a single-cell dual-omics approach.

Another prospective approach would be to find a possibility of physically separating the (m)RNA from the DNA without the use of polyA tails. This could be accomplished by dividing the samples into two portions after cell lysis, which are then treated differently, based on which fraction is supposed to be extracted, e.g., either with DNase to recover RNA or RNase to recover DNA. As microbial single-cell approaches in general already struggle with the minute amounts of DNA and RNA in a cell (Table 7.1), wasting half of the RNA or DNA fraction might theoretically be possible, however, difficult to achieve experimentally. Nevertheless, this approach would be suitable for MDM.

7.2.2 Genomics & Proteomics

In 2021, Mimitou et al. combined single-cell genomics and proteomics in eukaryotes for the first time (Mimitou et al. 2021). This approach relies on the detection of cell surface markers with oligo-conjugated antibodies, making it unsuitable for acquiring novel protein information or generally assessing MDM. One potential approach for prokaryotes could be by physically splitting the sample after cell lysis into two parts and processing each differently according to the target analyte (DNA or protein). However, the challenge of low analyte concentrations persists (Table 7.1). Another strategy could involve a slightly modified version of CITE-seq (Stoeckius et al. 2017) (see Sect. 7.2.3), leveraging barcoding and targeted protein linking. For this, the DNA could be immediately subjected to amplification, circumventing the use of polyA tails. As CITE-seq targets specific proteins through antibodies as well, it would not be suitable for elucidating FDM. Additionally, the development of such a dual-omics approach for prokaryotes would face similar challenges to those encountered in the respective mono-omics, particularly regarding the detection limits of MS instruments, as previously mentioned (see Sects. 7.1 and 7.2.3).

7.2.3 Transcriptomics & Proteomics

The prokaryotic-specific polycistronic messenger RNAs (Table 7.1) are a major challenge for a dual approach integrating single-cell transcriptomics (SCT) and single-cell proteomics (SCP) in prokaryotes. Because multiple genes can share the same mRNA molecule, the amount of transcripts in a cell may not accurately reflect the expression of individual genes. Additionally, the translational efficiency of each cistron within a polycistronic mRNA can vary, such that a high abundance of the transcript does not necessarily correlate with the abundance of its translated proteins. Furthermore, proteins derived from polycistronic mRNAs may potentially share similar MS signatures, which could further complicate their accurate quantification using MS. For eukaryotes, this dual-omics approach has been achieved by collecting protein information using predetermined antibodies to tag proteins of specific cell phenotypes (Frei et al. 2016; Genshaft et al. 2016; Mimitou et al. 2019). Unfortunately, the utilization of predetermined antibodies does not allow for acquiring novel protein information, making this approach only suitable for application on characterized organisms and proteins.

One single-cell dual-omics approach of transcriptomics and proteomics in eukaryotes is CITE-seq (see Sec. 1.2.2). It uses antibodies conjugated with a polyA tail to bind the mRNA, as well as a unique oligonucleotide barcode that binds to the proteins of interest at the same time. The analyzed cell is encapsulated in droplets together with the antibody–oligo complexes and the beads with oligonucleotides containing a unique molecular identifier (UMI) and cell-specific barcode. The mRNA of the cell and the antibody–oligo complexes are captured through beads by their polyA tail. The captured RNA is then reverse transcribed to cDNA, and the antibody-derived oligonucleotides are amplified. The unique barcodes ensure that the RNA and protein information from each cell is retained, and the amplified products are then sequenced (Stoeckius et al. 2017). REAP-seq follows the same principle but differs in the way the oligonucleotide barcode is conjugated to the antibody (Peterson et al. 2017). Due to their reliance on polyA tails, these methods are not suitable for a prokaryotic dual-omics approach.

In 2016, Darmanis et al. used oligonucleotide–antibody conjugates to a protein currently hindered by the presence of polycistronic mRNAs, as already mentioned before. Additionally, the method relies on antibodies that bind to specific proteins as well. That makes this method not suitable for universal protein identification and elucidation of MDM or FDM.

Ultimately, for combining transcriptomics and proteomics on one single cell in prokaryotes, it is first necessary to address the challenges associated with each individual mono-omics approach. Established workflows for SCT and SCP diverge after cell lysis. For SCT, the RNA is reverse transcribed and sequenced (Fig. 7.1). The SCP protocol published by Végvári et al. continues with protein denaturation, digestion with trypsin and TEAB, TMT labeling, and MS against protein databases (Fig. 7.1). Comparing the experimental setups, a combined workflow appears feasible by splitting the sample after cell lysis and processing the RNA and protein parts accordingly. For instance, the RNA fraction could undergo gDNA removal and reverse transcription, while the protein fraction could be subjected to digestion and labeling (Fig. 7.1). The waste of half of the proteins of the cell is nevertheless problematic due to the low detection limits of MS. Maximizing protein capture is therefore critical. In the SCP protocol by Végvári et al., the limitation of low protein concentrations was circumvented by using a fast-growing organism and hence a high abundance of ribosomal proteins, yielding protein levels sufficient for SCoPE-MS, the current gold standard of eukaryotic single-cell proteomics (Budnik et al. 2018). Unfortunately, most prokaryotes exhibit slow growth rates, resulting in fewer ribosomes and reduced protein abundance (Neidhardt and Magasanik 1960; Scott and Hwa 2011). Consequently, the current SCP protocol is not applicable to prokaryotes in general.

7.2.4 Proteomics and Metabolomics

Wu et al. recently published a method (scPMA) integrating SCP and SCM in eukaryotes for the first time. The process begins with the isolation and capture of single tumor cells, followed by cell lysis, protein reduction, alkylation, enzymatic digestion of proteins, termination of digestion, and LC–MS/MS analysis. The authors developed a novel dual-zone detection approach that applies distinct MS detection modes within one single sample injection cycle. In the first minutes of the LC–MS/MS run, MS parameters optimized for peptide detection are used for protein analysis, after which the parameters are switched to those optimized for metabolite profiling. Identification was then achieved by matching the obtained MS/MS spectra to known reference data (Wu et al. 2024). Mass spectrometry might therefore provide a unified method for the integration of single-cell proteomics and metabolomics in prokaryotes. Additionally, MS is a label-free approach, harboring the potential for application on unknown bacteria and archaea as well as metabolic pathways. Adapting this method for prokaryotes would present the same challenges regarding the detection limits of MS instruments as previously mentioned (see Sects. 7.1 and 7.2.3). Additionally, this method is a "one-shot method", making other downstream experiments unfeasible. A workflow of genomics or transcriptomics in combination with metabolomics has not yet been archived for single-cells, neither in eukaryotes nor in prokaryotes.

7.3 Conclusion and Outlook

In conclusion, each single-cell mono-omics assesses a different aspect of a living organism, from basic genome information and gene expression levels to functionality and whole metabolism via proteins and metabolites. Nevertheless, to fully understand single-cell biology in a complete and comprehensive way, multiple "omics" technologies have to be combined and applied simultaneously to one individual cell. Currently, only a few dual single-cell omics approaches for eukaryotic cells exist to date; some might be feasible for adaptation in prokaryotes, especially for a dual SCG and SCT approach. However, no method combining three or all four omics technologies has been developed yet, and high-throughput applications are quite far from realization. In general, the amounts of analytes present in the current state of a cell pose a significant challenge, especially for prokaryotes. Detection limitations of instruments are currently not sensitive enough to correctly identify molecules on a single-cell level, and advances in reducing input material and enhancing detection sensitivity need to be made. In particular, reduced molecular fragmentation in MS-based techniques is essential for advancing spectrometry-based single-cell techniques, but also reducing input amounts of nucleic acids for sequencing. The differences in the experimental workflows further impede the integration into a single-cell multi-omics workflow. The extensive expertise and high costs associated with single-cell studies also pose a significant barrier in this research area. These studies require specialized instruments, strong computational resources, as well as single-cell expertise (such as single-cell sorting, cell lysis, and labeling methods). As a result, single-cell analysis is still performed by only a few laboratories with the necessary equipment and specialization.

The future of single-cell multi-omics technology development might be driven by leveraging artificial intelligence (AI). The rapid advancements in AI and machine learning in recent years have enabled analyzing large, complex datasets to make accurate predictions from noisy or incomplete data. In multi-omics studies, AI-driven pattern recognition from public databases could infer the composition and abundance of proteins, metabolites, and other cellular components without requiring direct measurements of each molecule. The application of AI will also help with the integration of different mono-omics data, enabling a joint analysis to create more accurate and predictive models of cellular function and behavior. This also means that costs of bioinformatic analysis of multi-omics and multi-layered data will scale up with the number of parameters and data points. In order to address this problem, researchers will require dedicated servers with high computing power. Depending on the type of multi-omics, reference databases have to be added and linked with each other. However, technical progress in this field will hopefully soon revolutionize microbiology, just as the development of meta-omics technologies has done in the past.

Competing Interests The authors declare no conflict of interest.

Funding The present contribution is supported by the Karlsruhe Institute of Technology, Institute for Biological Interfaces 5, Department of Microbial Genetics & Biotechnology. Further funding was provided by The Helmholtz Association of German Research Centers through its research program "PoF IV".

References

Acuña-Amador L, Primot A, Cadieu E et al (2018) Genomic repeats, misassembly and reannotation: a case study with long-read resequencing of *Porphyromonas gingivalis* reference strains. BMC Genomics 19:1–24. https://doi.org/10.1186/S12864-017-4429-4

Ahmad R, Budnik B (2023) A review of the current state of single-cell proteomics and future perspective. Anal Bioanal Chem 415:6889–6899. https://doi.org/10.1007/s00216-023-04759-8

Alberts B, Heald R, Johnson A et al (2022) Molecular biology of the cell. W. W. Norton & Company

Amann R, Fuchs BM (2008) Single-cell identification in microbial communities by improved fluorescence in situ hybridization techniques. Nat Rev Microbiol 6:339–348. https://doi.org/10.1038/nrmicro1888

Amantonico A, Urban PL, Zenobi R (2010) Analytical techniques for single-cell metabolomics: state of the art and trends. Anal Bioanal Chem 398:2493–2504. https://doi.org/10.1007/S00216-010-3850-1

Anand H, Balasundaram B, Pandit AB, Harrison STL (2007) The effect of chemical pretreatment combined with mechanical disruption on the extent of disruption and release of intracellular protein from *E. coli*. Biochem Eng J 35:166–173. https://doi.org/10.1016/j.bej.2007.01.011

Anderson JP, Angerer B, Loeb LA (2005) Incorporation of reporter-labeled nucleotides by DNA polymerases. Biotechniques 38:257–264. https://doi.org/10.2144/05382RR02

Ardern Z, Chakraborty S, Lenk F, Kaster AK (2023) Elucidating the functional roles of prokaryotic proteins using big data and artificial intelligence. FEMS Microbiol Rev 47. https://doi.org/10.1093/FEMSRE/FUAD003

Baysoy A, Bai Z, Satija R, Fan R (2023) The technological landscape and applications of single-cell multi-omics. Nat Rev Mol Cell Biol 24:695–713. https://doi.org/10.1038/s41580-023-00615-w

Beattie M, Jones OAH (2023) Rate of advancement of detection limits in mass spectrometry: is there a Moore's law of mass spec? Mass Spectrom 12:118. https://doi.org/10.5702/MASSSPECTROMETRY.A0118

Blainey PC (2013) The future is now: single-cell genomics of bacteria and archaea. FEMS Microbiol Rev 37:407–427. https://doi.org/10.1111/1574-6976.12015

Blattman SB, Jiang W, Oikonomou P, Tavazoie S (2020) Prokaryotic single-cell RNA sequencing by in situ combinatorial indexing. Nat Microbiol 5:1192–1201. https://doi.org/10.1038/S41564-020-0729-6

Bonnin EA, Rizzoli SO (2020) Novel secondary ion mass spectrometry methods for the examination of metabolic effects at the cellular and subcellular levels. Front Behav Neurosci 14. https://doi.org/10.3389/FNBEH.2020.00124

Bowers RM, Gonzalez-Pena V, Wardhani K et al (2024) scMicrobe PTA: near complete genomes from single bacterial cells. bioRxiv:2024.01.30.577819. https://doi.org/10.1101/2024.01.30.577819

Brennan MA, Rosenthal AZ (2021) Single-cell RNA sequencing elucidates the structure and organization of microbial communities. Front Microbiol 12:713128. https://doi.org/10.3389/fmicb.2021.713128

Budnik B, Levy E, Harmange G, Slavov N (2018) SCoPE-MS: mass spectrometry of single mammalian cells quantifies proteome heterogeneity during cell differentiation. Genome Biol 19. https://doi.org/10.1186/S13059-018-1547-5

Carvalhais LC, Dennis PG, Tyson GW, Schenk PM (2012) Application of metatranscriptomics to soil environments. J Microbiol Methods 91:246–251

Chen F, Ye J, Chio C et al (2020) A simplified quick microbial genomic DNA extraction *via* freeze–thawing cycles. Mol Biol Rep 47:703–709. https://doi.org/10.1007/s11033-019-05176-w

Cooper GM (2000) The cell. 8:103–108. https://www.ncbi.nlm.nih.gov/books/NBK9839

Cui L, Yang K, Li HZ et al (2018) Functional single-cell approach to probing nitrogen-fixing bacteria in soil communities by resonance Raman spectroscopy with 15N2 labeling. Anal Chem 90:5082–5089. https://doi.org/10.1021/acs.analchem.7b05080

Dam HT, Vollmers J, Sobol MS et al (2020) Targeted cell sorting combined with single cell genomics captures low abundant microbial dark matter with higher sensitivity than metagenomics. Front Microbiol 11:511433. https://doi.org/10.3389/FMICB.2020.01377

Dekkers KF, Sayols-Baixeras S, Baldanzi G et al (2022) An online atlas of human plasma metabolite signatures of gut microbiome composition. Nat Commun 13:1–12. https://doi.org/10.1038/s41467-022-33050-0

Dey SS, Kester L, Spanjaard B et al (2015) Integrated genome and transcriptome sequencing of the same cell. Nat Biotechnol 33:285–289. https://doi.org/10.1038/NBT.3129

Dusny C (2022) Microfluidic single-cell analytics. Adv Biochem Eng Biotechnol 179:159–189. https://doi.org/10.1007/10_2020_134

Escudeiro P, Henry CS, Dias RPM (2022) Functional characterization of prokaryotic dark matter: the road so far and what lies ahead. Curr Res Microb Sci 3:100159. https://doi.org/10.1016/J.CRMICR.2022.100159

Frei AP, Bava FA, Zunder ER et al (2016) Highly multiplexed simultaneous detection of RNAs and proteins in single cells. Nat Methods 13:269–275. https://doi.org/10.1038/NMETH.3742

Gaisser KD, Skloss SN, Brettner LM et al (2024) High-throughput single-cell transcriptomics of bacteria using combinatorial barcoding. Nat Protoc 2024:1–37. https://doi.org/10.1038/s41596-024-01007-w

Galperin MY, Koonin EV (2004) 'Conserved hypothetical' proteins: prioritization of targets for experimental study. Nucleic Acids Res 32:5452–5463. https://doi.org/10.1093/NAR/GKH885

Genshaft AS, Li S, Gallant CJ et al (2016) Multiplexed, targeted profiling of single-cell proteomes and transcriptomes in a single reaction. Genome Biol 17:1–15. https://doi.org/10.1186/s13059-016-1045-6

Ghatak S, King ZA, Sastry A, Palsson BO (2019) The y-ome defines the 35% of *Escherichia coli* genes that lack experimental evidence of function. Nucleic Acids Res 47:2446–2454. https://doi.org/10.1093/NAR/GKZ030

Grassl N, Kulak NA, Pichler G et al (2016) Ultra-deep and quantitative saliva proteome reveals dynamics of the oral microbiome. Genome Med 8. https://doi.org/10.1186/S13073-016-0293-0

Gregory TR (2005) Chapter 1 – Genome size evolution in animals. In: Gregory TR (ed) The evolution of the genome. Academic, Burlington, pp 3–87

Gross A, Schöndube J, Niekrawitz S et al (2013) Single-cell printer: automated, on demand, and label free. J Lab Autom 18:504–518. https://doi.org/10.1177/2211068213497204

Gross A, Schöndube J, Zimmermann S et al (2015) Technologies for single-cell isolation. Int J Mol Sci 16:16897–16919

Guo S, Zhang C, Le A (2021) The limitless applications of single-cell metabolomics. Curr Opin Biotechnol 71:115–122. https://doi.org/10.1016/J.COPBIO.2021.07.015

Han L, Zi X, Garmire LX et al (2014) Co-detection and sequencing of genes and transcripts from the same single cells facilitated by a microfluidics platform. Sci Rep 4:1–9. https://doi.org/10.1038/srep06485

Haque A, Engel J, Teichmann SA, Lönnberg T (2017) A practical guide to single-cell RNA-sequencing for biomedical research and clinical applications. Genome Med 9:1–12. https://doi.org/10.1186/s13073-017-0467-4

Hatzenpichler R, Krukenberg V, Spietz RL, Jay ZJ (2020) Next-generation physiology approaches to study microbiome function at single cell level. Nat Rev Microbiol 18:241–256

Homberger C, Hayward RJ, Barquist L, Vogel J (2023) Improved bacterial single-cell RNA-Seq through automated MATQ-Seq and Cas9-based removal of rRNA reads. MBio 14. https://doi.org/10.1128/MBIO.03557-22

Hong S, Pawel GT, Pei R, Lu Y (2021) Recent progress in developing fluorescent probes for imaging cell metabolites. Biomed Mater 16. https://doi.org/10.1088/1748-605X/ABFD11

Huang WE, Griffiths RI, Thompson IP et al (2004) Raman microscopic analysis of single microbial cells. Anal Chem 76:4452–4458. https://doi.org/10.1021/ac049753k

Hugenholtz P, Tyson GW (2008) Metagenomics. Nature 455:481–483. https://doi.org/10.1038/455481a

Imdahl F, Vafadarnejad E, Homberger C et al (2020) Single-cell RNA-sequencing reports growth-condition-specific global transcriptomes of individual bacteria. Nat Microbiol 5:1202–1206. https://doi.org/10.1038/s41564-020-0774-1

Islam MS, Aryasomayajula A, Selvaganapathy PR (2017) A review on macroscale and microscale cell lysis methods. Micromachines (Basel) 8. https://doi.org/10.3390/MI8030083

Jiao JY, Liu L, Hua ZS et al (2021) Microbial dark matter coming to light: challenges and opportunities. Natl Sci Rev 8:2021. https://doi.org/10.1093/NSR/NWAA280

Kang Y, Norris MH, Zarzycki-Siek J et al (2011) Transcript amplification from single bacterium for transcriptome analysis. Genome Res 21:925–935. https://doi.org/10.1101/GR.116103.110

Kaster AK, Sobol MS (2020) Microbial single-cell omics: the crux of the matter. Appl Microbiol Biotechnol 104:8209–8220. https://doi.org/10.1007/S00253-020-10844-0

Kuchina A, Brettner LM, Paleologu L et al (2021) Microbial single-cell RNA sequencing by split-pool barcoding. Science 371. https://doi.org/10.1126/SCIENCE.ABA5257

Kumar VBN, Guo S, Bocklitz T et al (2016) Demonstration of carbon catabolite repression in naphthalene degrading soil bacteria via Raman spectroscopy based stable isotope probing. Anal Chem 88:7574–7582. https://doi.org/10.1021/acs.analchem.6b01046

Lasken RS, McLean JS (2014) Recent advances in genomic DNA sequencing of microbial species from single cells. Nat Rev Genet 15:577–584. https://doi.org/10.1038/NRG3785

Lasken RS, Stockwell TB (2007) Mechanism of chimera formation during the multiple displacement amplification reaction. BMC Biotechnol 7:1–11. https://doi.org/10.1186/1472-6750-7-19

Lee J, Young Hyeon D, Hwang D (2020) Single-cell multiomics: technologies and data analysis methods. Exp Mol Med 52:1428–1442. https://doi.org/10.1038/s12276-020-0420-2

Liu Y, Beyer A, Aebersold R (2016) On the dependency of cellular protein levels on mRNA abundance. Cell 165:535–550. https://doi.org/10.1016/J.CELL.2016.03.014

Lloyd KG, Steen AD, Ladau J et al (2018) Phylogenetically novel uncultured microbial cells dominate earth microbiomes. mSystems:3. https://doi.org/10.1128/msystems.00055-18

Lo H-Y, Wink K, Nitz H et al (2023) scMAR-Seq: a novel workflow for targeted single-cell genomics of microorganisms using radioactive labeling. mSystems 8. https://doi.org/10.1128/MSYSTEMS.00998-23

Lok C (2015) Mining the microbial dark matter. Nature 522:270–273. https://doi.org/10.1038/522270A

Macek B, Forchhammer K, Hardouin J et al (2019) Protein post-translational modifications in bacteria. Nat Rev Microbiol 17:651–664

Marshall IPG, Blainey PC, Spormann AM, Quake SR (2012) A single-cell genome for Thiovulum sp. Appl Environ Microbiol 78:8555–8563. https://doi.org/10.1128/AEM.02314-12

McDonald D, Price MN, Goodrich J et al (2012) An improved Greengenes taxonomy with explicit ranks for ecological and evolutionary analyses of bacteria and archaea. ISME J 6:610–618. https://doi.org/10.1038/ISMEJ.2011.139

Milo R, Phillips R (2015) Cell biology by the numbers. Garland Science, New York. https://doi.org/10.1201/9780429258770

Mimitou EP, Cheng A, Montalbano A et al (2019) Multiplexed detection of proteins, transcriptomes, clonotypes and CRISPR perturbations in single cells. Nat Methods 16:409–412. https://doi.org/10.1038/s41592-019-0392-0

Mimitou EP, Lareau CA, Chen KY et al (2021) Scalable, multimodal profiling of chromatin accessibility, gene expression and protein levels in single cells. Nat Biotechnol 39:1246–1258. https://doi.org/10.1038/s41587-021-00927-2

Mozzi F, Ortiz ME, Bleckwedel J et al (2013) Metabolomics as a tool for the comprehensive understanding of fermented and functional foods with lactic acid bacteria. Food Res Int 54:1152–1161. https://doi.org/10.1016/J.FOODRES.2012.11.010

Münch JM, Sobol MS, Brors B, Kaster AK (2023) Single-cell transcriptomics and data analyses for prokaryotes—past, present and future concepts. Adv Appl Microbiol 123:1–39. https://doi.org/10.1016/BS.AAMBS.2023.04.002

Natarajan A, Srienc F (1999) Dynamics of glucose uptake by single *Escherichia coli* cells. Metab Eng 1:320–333. https://doi.org/10.1006/MBEN.1999.0125

Neidhardt FC, Magasanik B (1960) Studies on the role of ribonucleic acid in the growth of bacteria. Biochim Biophys Acta 42:99–116. https://doi.org/10.1016/0006-3002(60)90757-5

Nikolic N, Schreiber F, Dal Co A et al (2017) Cell-to-cell variation and specialization in sugar metabolism in clonal bacterial populations. PLoS Genet 13:e1007122. https://doi.org/10.1371/JOURNAL.PGEN.1007122

Ojala T, Häkkinen AE, Kankuri E, Kankainen M (2023) Current concepts, advances, and challenges in deciphering the human microbiota with metatranscriptomics. Trends Genet 39:686–702

Paige JS, Nguyen-Duc T, Song W, Jaffrey SR (2012) Fluorescence imaging of cellular metabolites with RNA. Science 335:1194. https://doi.org/10.1126/SCIENCE.1218298

Peterson VM, Zhang KX, Kumar N et al (2017) Multiplexed quantification of proteins and transcripts in single cells. Nat Biotechnol 35:936–939. https://doi.org/10.1038/NBT.3973

Picelli S, Faridani OR, Björklund ÅK, et al (2014) Full-length RNA-seq from single cells using smart-seq2. Nat Protoc 9:171–181. doi: https://doi.org/10.1038/nprot.2014.006

Podar M, Abulencia CB, Walcher M et al (2007) Targeted access to the genomes of low-abundance organisms in complex microbial communities. Appl Environ Microbiol 73:3205. https://doi.org/10.1128/AEM.02985-06

Pratscher J, Vollmers J, Wiegand S et al (2018) Unravelling the identity, metabolic potential and global biogeography of the atmospheric methane-oxidizing upland soil cluster α. Environ Microbiol 20:1016–1029. https://doi.org/10.1111/1462-2920.14036

Premasiri WR, Lee JC, Sauer-Budge A et al (2016) The biochemical origins of the surface-enhanced Raman spectra of bacteria: a metabolomics profiling by SERS. Anal Bioanal Chem 408:4631–4647. https://doi.org/10.1007/S00216-016-9540-X

Raghunathan A, Ferguson HR, Bornarth CJ et al (2005) Genomic DNA amplification from a single bacterium. Appl Environ Microbiol 71:3342–3347. https://doi.org/10.1128/AEM.71.6.3342-3347.2005

Riba J, Gleichmann T, Zimmermann S et al (2016) Label-free isolation and deposition of single bacterial cells from heterogeneous samples for clonal culturing. Sci Rep 6. https://doi.org/10.1038/SREP32837

Rinke C, Schwientek P, Sczyrba A et al (2013) Insights into the phylogeny and coding potential of microbial dark matter. Nature 499:431–437. https://doi.org/10.1038/NATURE12352

Rinke C, Lee J, Nath N et al (2014) Obtaining genomes from uncultivated environmental microorganisms using FACS–based single-cell genomics. Nat Protoc 9:1038–1048. https://doi.org/10.1038/nprot.2014.067

Robinson SL, Piel J, Sunagawa S (2021) A roadmap for metagenomic enzyme discovery. Nat Prod Rep 38:1994–2023. https://doi.org/10.1039/D1NP00006C

Rosenow C, Saxena RM, Durst M, Gingeras TR (2001) Prokaryotic RNA preparation methods useful for high density array analysis: comparison of two approaches. Nucleic Acids Res 29. https://doi.org/10.1093/NAR/29.22.E112

Rutledge PJ, Challis GL (2015) Discovery of microbial natural products by activation of silent biosynthetic gene clusters. Nat Rev Microbiol 13:509–523

Sabina J, Leamon JH (2015) Bias in whole genome amplification: causes and considerations. Methods Mol Biol 1347:15–41. https://doi.org/10.1007/978-1-4939-2990-0_2

Schena M, Shalon D, Davis RW, Brown PO (1995) Quantitative monitoring of gene expression patterns with a complementary DNA microarray. Science 270:467–470. https://doi.org/10.1126/SCIENCE.270.5235.467

Schmid A, Kortmann H, Dittrich PS, Blank LM (2010) Chemical and biological single cell analysis. Curr Opin Biotechnol 21:12–20. https://doi.org/10.1016/J.COPBIO.2010.01.007

Scott M, Hwa T (2011) Bacterial growth laws and their applications. Curr Opin Biotechnol 22:559–565. https://doi.org/10.1016/J.COPBIO.2011.04.014

Sheng K, Cao W, Niu Y et al (2017) Effective detection of variation in single-cell transcriptomes using MATQ-seq. Nat Methods 14:267–270. https://doi.org/10.1038/nmeth.4145

Silhavy TJ, Kahne D, Walker S (2010) The bacterial cell envelope. Cold Spring Harb Perspect Biol 2(5):a000414-a000414. https://doi.org/10.1101/cshperspect.a000414

Sobol MS, Sturm G, Kaster A-K (2024) Besser und günstiger: Volumenreduktion in der Einzelzellgenomik. BIOspektrum 30:33–36. https://doi.org/10.1007/S12268-024-2126-9/METRICS
Stepanauskas R, Sieracki ME (2007) Matching phylogeny and metabolism in the uncultured marine bacteria, one cell at a time. Proc Natl Acad Sci U S A 104:9052–9057. https://doi.org/10.1073/PNAS.0700496104
Stoeckius M, Hafemeister C, Stephenson W et al (2017) Simultaneous epitope and transcriptome measurement in single cells. Nat Methods 14:865–868. https://doi.org/10.1038/nmeth.4380
Sturm G, Mojarrad M, Kaster AK (2023) Targeted cell labeling and sorting of prokaryotes for cultivation and omics approaches. Microb Physiol 33:63–84. https://doi.org/10.1159/000532088
Tajik M, Baharfar M, Donald WA (2022) Single-cell mass spectrometry. Trends Biotechnol 40:1374–1392. https://doi.org/10.1016/J.TIBTECH.2022.04.004
Thomson B (2012) Driving high sensitivity in biomolecular MS. Genet Eng Biotechnol News 32:17. https://doi.org/10.1089/gen.32.20.08
Tourrière H, Chebli K, Tazi J (2002) mRNA degradation machines in eukaryotic cells
Van Den Bossche T, Arntzen M, Becher D et al (2021) The metaproteomics initiative: a coordinated approach for propelling the functional characterization of microbiomes. Microbiome 9
Vandereyken K, Sifrim A, Thienpont B, Voet T (2023) Methods and applications for single-cell and spatial multi-omics. Nat Rev Genet 24:494–515. https://doi.org/10.1038/s41576-023-00580-2
Végvári Á, Zhang X, Zubarev RA (2023) Toward single bacterium proteomics. J Am Soc Mass Spectrom 34:2098. https://doi.org/10.1021/JASMS.3C00242
Vincent AT (2024) Bacterial hypothetical proteins may be of functional interest. Front Bacteriol 3:1334712. https://doi.org/10.3389/FBRIO.2024.1334712
Vollmers J, Wiegand S, Kaster AK (2017) Comparing and evaluating metagenome assembly tools from a microbiologist's perspective - not only size matters! PLoS One 12:e0169662. https://doi.org/10.1371/JOURNAL.PONE.0169662
Vollmers J, Wiegand S, Lenk F, Kaster AK (2022) How clear is our current view on microbial dark matter? (Re-)assessing public MAG & SAG datasets with MDMcleaner. Nucleic Acids Res 50:e76–e76. https://doi.org/10.1093/NAR/GKAC294
Vollmers J, Cassal MC, Kaster A-K (2023) "Midi-metagenomics": a novel approach for cultivation independent microbial genome reconstruction from environmental samples. bioRxiv:2023.01.26.525644. https://doi.org/10.1101/2023.01.26.525644
Wang Y, Navin NE (2015) Advances and applications of single-cell sequencing technologies. Mol Cell 58:598–609. https://doi.org/10.1016/J.MOLCEL.2015.05.005
Wang J, Chen L, Chen Z, Zhang W (2015) RNA-seq based transcriptomic analysis of single bacterial cells. Integr Biol 7:1466–1476. https://doi.org/10.1039/C5IB00191A
Wang Y, Huang WE, Cui L, Wagner M (2016) Single cell stable isotope probing in microbiology using Raman microspectroscopy. Curr Opin Biotechnol 41:34–42. https://doi.org/10.1016/J.COPBIO.2016.04.018
Wang B, Lin AE, Yuan J et al (2023) Single-cell massively-parallel multiplexed microbial sequencing (M3-seq) identifies rare bacterial populations and profiles phage infection. Nat Microbiol 8:1846–1862. https://doi.org/10.1038/S41564-023-01462-3
Westermann AJ, Vogel J (2021) Cross-species RNA-seq for deciphering host–microbe interactions. Nat Rev Genet 22:361–378. https://doi.org/10.1038/s41576-021-00326-y
Wiegand S, Dam HT, Riba J et al (2021) Printing microbial dark matter: using single cell dispensing and genomics to investigate the patescibacteria/candidate phyla radiation. Front Microbiol 12:635506. https://doi.org/10.3389/FMICB.2021.635506
Wiegand S, Sobol M, Schnepp-Pesch LK et al (2023) Taxonomic re-classification and expansion of the phylum chloroflexota based on over 5000 genomes and metagenome-assembled genomes. Microorganisms 11:2612. https://doi.org/10.3390/MICROORGANISMS11102612
Wishart DS, Guo AC, Oler E et al (2022) HMDB 5.0: the human metabolome database for 2022. Nucleic Acids Res 50:D622–D631. https://doi.org/10.1093/NAR/GKAB1062
Woyke T, Doud DFR, Schulz F (2017) The trajectory of microbial single-cell sequencing. Nat Methods 14:1045–1054. https://doi.org/10.1038/nmeth.4469

Wu J, Xu QQ, Jiang YR et al (2024) One-shot single-cell proteome and metabolome analysis strategy for the same single cell. Anal Chem 96:5499–5508. https://doi.org/10.1021/acs.analchem.3c05659

Xu J, Zhu D, Ibrahim AD et al (2017) Raman deuterium isotope probing reveals microbial metabolism at the single-cell level. Anal Chem 89:13305–13312. https://doi.org/10.1021/ACS.ANALCHEM.7B03461

Yin Y, Jiang Y, Lam KWG et al (2019) High-throughput single-cell sequencing with linear amplification. Mol Cell 76:676–690.e10. https://doi.org/10.1016/J.MOLCEL.2019.08.002

Yu L, Wang X, Mu Q et al (2023) scONE-seq: a single-cell multi-omics method enables simultaneous dissection of phenotype and genotype heterogeneity from frozen tumors. Sci Adv 9. https://doi.org/10.1126/sciadv.abp8901

Zachariadis V, Cheng H, Andrews N, Enge M (2020) A highly scalable method for joint whole-genome sequencing and gene-expression profiling of single cells. Mol Cell 80:541–553.e5. https://doi.org/10.1016/J.MOLCEL.2020.09.025

Zenobi R (2013) Single-cell metabolomics: analytical and biological perspectives. Science 342. https://doi.org/10.1126/SCIENCE.1243259

Zhan L, Liu H, Hou Z et al (2023) Recent advances in single bacterium metabolic analysis techniques. TrAC Trends Anal Chem 163:117076. https://doi.org/10.1016/J.TRAC.2023.117076

Zhang L, Cui X, Schmitt K et al (1992) Whole genome amplification from a single cell: implications for genetic analysis. Proc Natl Acad Sci U S A 89:5847–5851. https://doi.org/10.1073/PNAS.89.13.5847

Zhang D, Qin Q, Qiao L (2023) Mass spectrometry profiling of single bacterial cells reveals metabolic regulation during antibiotics induced bacterial filamentation. Chin Chem Lett 34:107938. https://doi.org/10.1016/J.CCLET.2022.107938

Single-Cell Sorting to Assess Phage Sensitivity in Heterogeneous Bacterial Populations

8

Lucas Mora-Quilis and Pilar Domingo-Calap

Abstract

Bacterial populations exhibit significant heterogeneity, even among genetically identical members of the same population, due to changes in gene expression. Individual variants, under certain selective pressures including environmental changes, antimicrobials or phages, can influence the evolutionary trajectory of a bacterial population. Phages target specific bacterial surface structures, and bacterial strategies such as the loss or modification of these receptors can prevent phage attachment, often through reversible changes in gene expression. However, such modifications can also alter the susceptibility to other phages or antibiotics. Understanding how bacterial heterogeneity influences phage–host interactions is crucial for studying coevolution and adaptation. While the study of resistance mechanisms often involves the isolation of phage-resistant variants, the transient nature of some bacterial phenotypes poses a challenge. This study proposes a cell sorting-based method to separate individual cells from heterogeneous populations and test their phenotypes under different treatments. We illustrate the application of this protocol with a representative bacterial culture infected by a single phage, where resistant variants emerge and increase in frequency gaining sensitivity to another phage. The protocol allows for the estimation of resistant and sensitive cells to different treatments (e.g., phages), providing insight into the heterogeneity of bacterial cultures.

L. Mora-Quilis · P. Domingo-Calap (✉)
Institute for Integrative Systems Biology (I2SysBio), University of Valencia and Spanish National Research Council (CSIC), Valencia, Spain
e-mail: pilar.domingo@uv.es

C. Rinke, M. Džunková (eds.), *Single-Cell Omics in Microbiome Research*, The Microbiomes of Humans, Animals, Plants, and the Environment 7,
https://doi.org/10.1007/978-3-032-07527-7_8

Keywords

Bacterial heterogeneity · Phage resistance · Transient phenotypes · Single-cell sorting · Phage–host interactions

8.1 Introduction

The heterogeneity of bacterial populations can be highly variable, even between members of the same clone-derived population (Magdanova and Golyasnaya 2013; Weigel and Dersch 2018; Ruiz et al. 2020; Kümmerli and Frank 2023; Ryall et al. 2012). Changes in gene expression may promote the coexistence of phenotypically distinct variants that, under certain selective pressures, may influence the evolutionary trajectory of the population (van der Woude and Bäumler 2004; Bayliss 2009; Gefen and Balaban 2009). Recently, new approaches, such as single-cell omics, have allowed the study of individual cells. Particularly, single-cell whole transcriptome sequencing is a key approach to unravelling the gene expression profile of a particular variant (Davis and Isberg 2016; Homberger et al. 2022; Müller and Nebe-von-Caron 2010). However, correlating this result with a given phenotype can be challenging, especially if the transition between different phenotypes is rapid, making it difficult to study and isolate variants with a non-stable phenotype.

In natural populations, bacteria coexist with their viruses (bacteriophages or phages), modulating their structure and composition. In this context, population heterogeneity might be crucial for evading phage infection (Turkington et al. 2019; Bull et al. 2014; Kim and Ryu 2012). Phages rely on their ability to recognize specific structures on the bacterial surface, such as outer membrane proteins, lipopolysaccharides, or capsule polysaccharides, to initiate infection (Bertozzi Silva et al. 2016; Gordillo Altamirano and Barr 2021). Consequently, the loss or reduction of these receptors is an effective bacterial strategy to prevent phage attachment (Azam and Tanji 2019; Li et al. 2022; Adler et al. 2021). This can occur through genetic mutations or by changes in gene expression that allow bacterial populations to switch between phenotypes quickly and reversibly, avoiding the potential costs associated with irreversible mutations. Interestingly, such modifications in cell surface receptors may confer sensitivity to previously non-infective phages or to chemical compounds such as antibiotics (Mangalea and Duerkop 2020; Cota et al. 2015; Chan et al. 2016).

Therefore, exploring how this heterogeneity can prevent or modify phage–host interactions is essential for understanding their coevolution and adaptation to different environments. One of the most common approaches to studying resistance mechanisms is the isolation of phage-resistant cultures. This process begins by challenging a bacterial culture with phages to promote the emergence of resistant variants. This is followed by several rounds of culture washing to reduce the presence of infecting particles. To ensure complete phage elimination, additional colony-to-colony passages are often performed. This is a straightforward method for isolating resistant mutant cultures due to their phenotypic stability. However, it may be inadequate for identifying variants that acquire resistance through changes in gene expression, as these phenotypes can rapidly revert to the wild-type phenotype once the selective pressure is removed. This limitation highlights the need for approaches capable of analysing transient phenotypes before reversion occurs. Single-cell methods allow immediate testing of individual cell phenotypes, without allowing time for growth or phenotypic reversion, thereby providing a real-time snapshot of the population's phenotypic diversity.

Here, we propose a method based on cell sorting to separate individual cells, which can be tested against different treatments. This approach allows for the estimation of the proportion of cells sensitive to each treatment within heterogeneous populations. We illustrate the application of this protocol using a representative bacterial culture that can be specifically infected by a single phage. Following phage infection, resistant variants emerge and rapidly increase in frequency within the population. Both original and resistant cultures are analysed by flow cytometry, followed by single-cell sorting, in which individual cells are exposed to different treatments to distinguish between phenotypes. After incubation, the number of wells with bacterial growth for each treatment is counted to estimate the proportion of resistant and sensitive cells in each culture, providing a measure of population heterogeneity.

The advantages and limitations of this protocol are described in **Notes 1** and **2**, respectively.

8.2 Materials

8.2.1 Media and Buffers

1. Any bacterial medium that supports the growth of the target bacteria can be used. Here, we describe the composition of a widely used bacterial medium, lysogeny broth (LB), supplemented with $CaCl_2$ to enhance phage adsorption:

   - NaCl (Labkem, SOCH-00A-500): 1 g/L.
   - Yeast extract (Labkem, YEEX-1 MB-500): 5 g/L.
   - Triptone (Labkem, TRYN-00P-500): 1 g/L.
   - $CaCl_2$ (Fisher Scientific, 10,161,800: 0.42 g/L.
   - Bacteriological agar (Labkem, BAC ISO-4833):

   (i) LB agar plates: 15 g/L.
   (ii) Semi-solid LB agar: 7.5 g/L.

   Dissolve all components in Milli-Q water. Sterilize by autoclaving at 120 °C for 20 min.
2. Phosphate-buffered saline (PBS) (Gibco, 14,040,133) 1x in Milli-Q water. Sterilize by autoclaving at 120 °C for 20 min, and filter through 0.22 μm to avoid the presence of large particles that could interfere with the fluorescence-activated cell sorting (FACS).

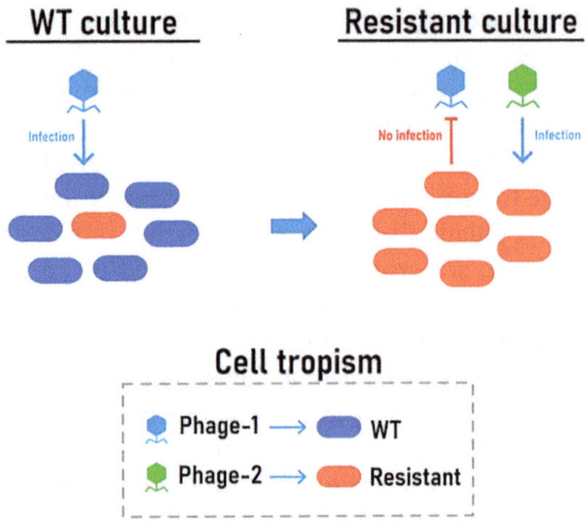

Fig. 8.1 Phage host tropism. Phage-1 infects specifically wild-type culture (WT) bacteria. The emergence of Phage-1-resistant variants allows susceptibility to Phage-2

3. 13 mL screw-cap tubes (SARSTEDT, 60.540.500) to dispense the semi-solid LB agar, maintained at 55 °C to prevent solidification.
4. Petri dishes 90 mm (Thermo Scientific, 101VR20) to dispense the LB agar.

8.2.2 Bacteria and Phage Strains

Any bacterial culture with its corresponding infecting-phage may be used for these experiments. As an example, we describe a representative system (Fig. 8.1) (*see* **Note 3**).

Bacteria:

- Wild-type culture (WT): Sensitive to Phage-1.
- Resistant culture: Emerges after WT infection with Phage-1 and becomes sensitive to Phage-2.

Phage:

- Phage-1: Infects the WT culture.
- Phage-2: Infects the resistant culture.

8.2.3 Preparation of Bacteria and Phage Suspensions for FACS

1. 1.5 mL microcentrifuge tubes (Eppendorf, 0030123611).
2. 96-well plate, flat transparent base (SARSTEDT, 82.1581.001).
3. Disposable reagent reservoir 45 mL (Astik's, RTIP-E51–250).
4. 500 mL glass Erlenmeyer flask.
5. 500 glass bottles (Fisherbrand, 15,476,113).

6. 200 μL and 1000 μL pipettes with corresponding sterile tips.
7. 300 μL or 1000 μL multichannel automated pipette (Eppendorf Xplorer Plus).
8. Incubator shaker (HT Infors Ecotron).
9. Sterile filter units (0.22 μm and 0.45 μm) (Thermo Scientific, 595–4520 and 295–4545).
10. Laboratory vacuum pump (Integra Biosciences, 158,300).
11. Centrifuge (Eppendorf 5810R).

8.2.4 Bacteria Single-Cell Sorting and Counting

1. BD FACS Aria fusion (BD Biosciences), with the following parameters for single-cell sorting:
    - Nozzle: 70 μm; pressure: 70 psi.
    - Sorting precision mode: Single-cell.
    - BD FACSFlow sheath fluid was used to clean the cytometer between samples.
2. Zip-lock bags or Parafilm.

8.3 Methods

8.3.1 Phage Suspensions

The following requirements must be met for phage suspensions:

- Ensure each phage suspension has a minimum concentration of $\geq 10^8$ plaque-forming units (PFU)/mL.
- Prepare sufficient volume of each phage suspension to fill all desired wells. A minimum of five 96-well plates per treatment is recommended (total of 480 wells/treatment).
- Filter the phage suspensions through 0.22 μm filters to remove bacterial debris.

An illustrative example for amplifying Phage-1 and Phage-2 is shown. Incubation times and temperatures may vary depending on the host and phage used:

1. In a sterile glass flask, mix 1 mL of a Phage-1 lysate containing 10^9 PFU/mL with 1 mL of stationary-phage WT bacteria in a final volume of 100 mL of LB (see **Note 4**). Repeat the process for Phage-2 amplification, using its respective bacterial host.
2. Incubate the flasks at 37 °C with agitation for 3 h.
3. After incubation, check for lysis: Flasks with phage should be transparent (indicating lysis), while negative control flasks should be turbid (indicating bacterial growth).

4. Filter the phage suspensions through 0.45 μm filters using a vacuum pump and collect the filtrate in a sterile glass bottle. Repeat the process filtering through 0.22 μm to ensure complete removal of the bacteria.
5. Ensure that the phage concentration is at the desired level (≥10^8 PFU/mL). Use the corresponding host for each phage:
   (a) Perform ten-fold serial dilutions of the amplified phage suspension.
   (b) Inoculate 10 μL of each phage dilution with 200 μL of stationary-phage host bacteria (approx. 10^9 colony-forming units (CFU)/mL) and mix with 3.5 mL of semi-solid LB agar containing $CaCl_2$.
   (c) Vortex briefly, then pour the mixture onto LB agar plates having $CaCl_2$.
   (d) Incubate the plates overnight at 37 °C.
   (e) Count the resulting PFU to determine the phage concentration.
6. Dispense 100 μL/well of each phage suspension in 96-well plates separately. Dispense 100 μL/well of LB medium supplemented with $CaCl_2$ to the other 96-well plates. A minimum of five 96-well plates per treatment is recommended to ensure an adequate number of counted wells. Use an automated multichannel pipette to streamline the process.
7. Close the lids and store the plates at 4 °C until FACS.

8.3.2 Bacterial Suspensions for FACS

1. Transfer 500 μL of the desired bacterial culture (approx. 10^9 CFU/mL) into a 1.5 mL tube (e.g., WT and resistant cultures).
2. Wash the cultures with PBS 1x. Centrifuge at 8000 *x g* for 3 minutes at 4 °C (*see* **Note 5**). Discard the supernatant and add 500 μL of 1x PBS. Repeat the washing step five times (*see* **Note 6**).
3. Dilute the bacterial suspension to a final volume of 5 mL in 0.22 μm-filtered 1x PBS to achieve an optimal concentration for flow cytometry (approx. 10^5 CFU/mL).
4. Store the tubes at 4 °C.

8.3.3 Bacteria Single-Cell Sorting by Flow Cytometry

1. Select the target population based solely on forward scatter area (FSC-A) and side scatter area (SSC-A) parameters (Fig. 8.2a).
2. To avoid doublets or cell aggregates, define the gate for individual cells based on SSC-A and side scatter height (SSC-H) (Fig. 8.2b).
3. Pass the bacterial cultures through the cytometer to sort individual cells in the 96-wells containing the different treatments (Fig. 8.3a).
   - First, pass the cultures that do not contain phages (e.g., WT) and sort single-cells into 96-well plates containing the different treatments.
   - Then, proceed with cultures that contain phages (e.g., resistant culture).

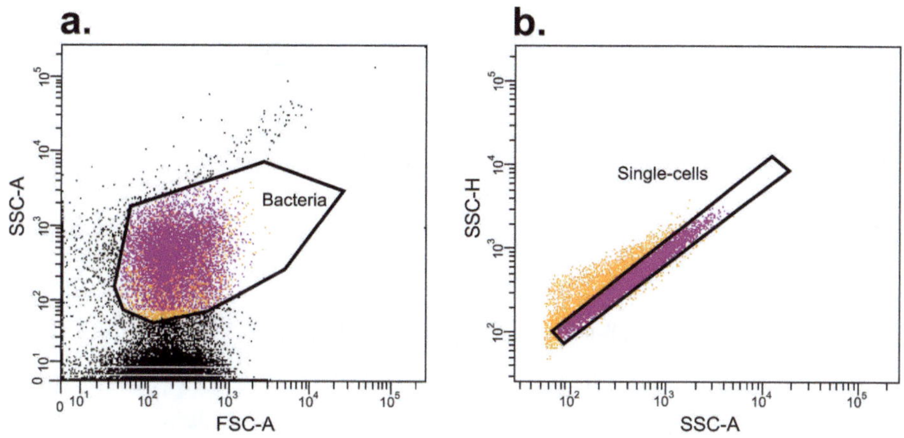

Fig. 8.2 Example of the flow cytometry scatter for the bacteria to be sorted. (**a**) The bacterial population might be selected based on forward scatter area (FSC-A) and side scatter area (SSC-A) parameters. (**b**) Avoid the selection of doublets or cell aggregates based on SSC-A and side scatter height (SSC-H)

To prevent cross-contamination between bacterial cultures, the cytometer system should be disinfected using FACSFlow Sheath fluid between samples.

8.3.4 Incubation and Counting Number of Wells with Bacterial Growth

1. Seal the plates with parafilm or place them in zip-lock bags to prevent desiccation.
2. Incubate plates at 37 °C with mild shaking for 48 h to allow single cells to grow (*see* **Note 7**).
3. Count the number of wells with bacterial growth for each treatment (Fig. 8.3b).
4. Consider the number of wells with growth in the LB-only control plates as the maximum possible growth, accounting for potential flow-cytometry error and the chance of sampling nonviable cells. Calculate growth in the other treatments as a percentage of this maximum.

8.3.5 Recommended Controls Prior to Performing the Experiments

- Test the isolation of single-cells via cytometry: This control ensures that the cytometer effectively separates individual cells. To perform this test, pass the selected culture through the cytometer and separate individual cells into wells containing medium (e.g., LB). It is recommended to use at least one 96-well plate for this test. Immediately after cell sorting, plate, and spread the contents of

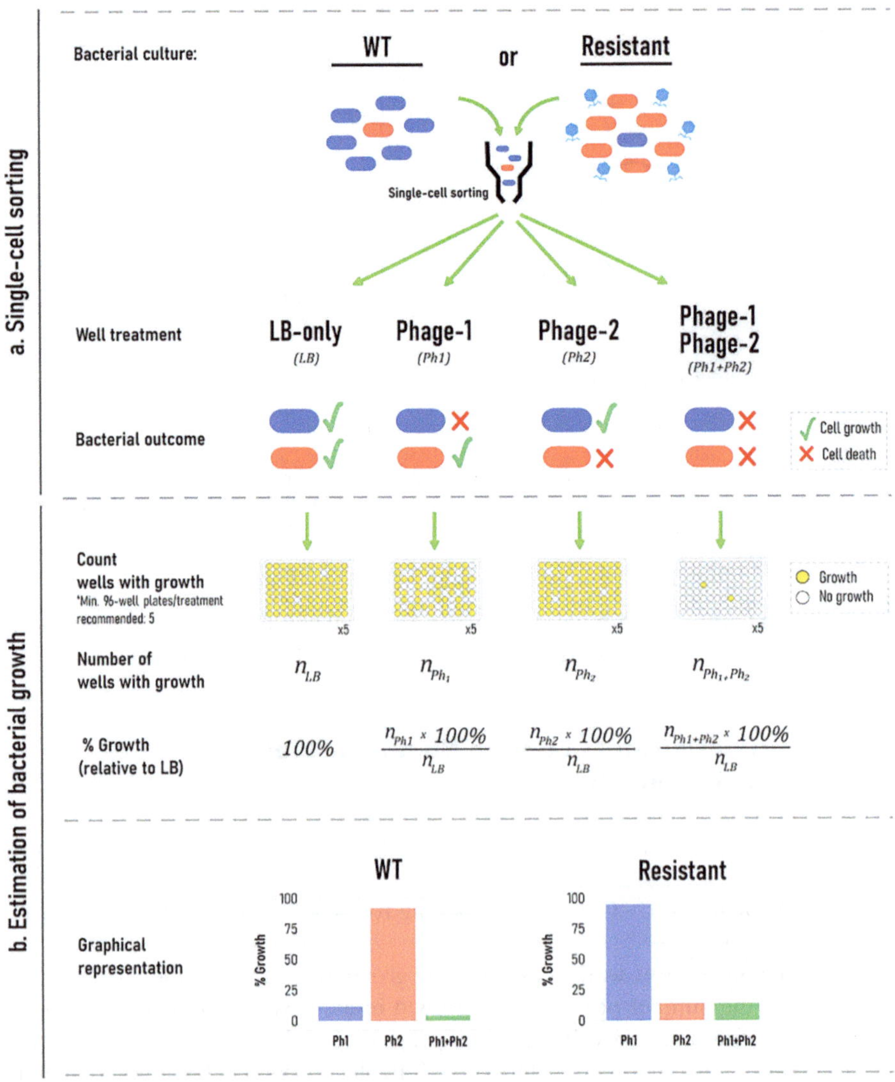

Fig. 8.3 Single-cell sorting protocol to estimate bacterial population heterogeneity. (**a**) Schematic representation of the bacterial single-cell sorting for the WT and the resistant cultures into wells containing different treatments, indicating cell growth or death. The WT variants (blue) are susceptible to Phage-1, while the resistant variants (red) are susceptible to Phage-2. Note that, through cytometry, cells from a resistant culture can be separated from residual Phage-1 present in the media. (**b**) Count of wells with bacterial growth in plates containing different treatments. A minimum of five 96-well plates per treatment (a total of 480 wells) is recommended to ensure an adequate number of counted wells. The number of wells with bacterial growth in each treatment is expressed relative to the LB-only control, which is considered 100%

each well onto separate agar plates. Incubate the plates and assess them for the presence of individual bacterial colonies. Some plates may have no colonies due to cytometer error or the presence of nonviable cells. However, it is not expected to observe two or more colonies from a single well.

- Test the complete removal of phages from the surrounding environment: If working with bacterial cultures that may contain phage particles (e.g., resistant culture), assess whether cell sorting separates single bacterial cells without carrying phages from the surrounding environment. Ensure a low phage concentration in the culture by performing washing steps prior to cell sorting (see Sect. 8.3.2). To validate this, pass the phage-containing culture through the cytometer and separate single cells into wells containing noninfected sensitive bacteria (e.g., WT) at a low concentration (e.g., 10^4–10^5 CFU/mL). Incubate the plates overnight, then transfer the well contents to 1.5 mL tubes. Centrifuge to pellet the bacterial culture, and plate the supernatant to test for the presence of phages. If phages had passed with the bacterial cell, they would replicate in the presence of the sensitive bacteria in the well, resulting in detectable phages in the supernatant.

8.4 Notes

1. The advantages of this protocol include:
   - Individual cells can be exposed not only to phages but also to other compounds of interest, such as antibiotics.
   - It enables the isolation of phenotypically transient bacteria and allows for rapid exposure to treatments (e.g., phages or antibiotics), preventing the bacteria from reverting to the wild-type phenotype.
   - In the context of phage resistance, this protocol isolates resistant variants from excess phages, which are typically difficult to remove through washing alone.
   - Unlike traditional colony picking for resistant variant isolation, flow cytometry can prevent the isolation of cell aggregates, avoiding doublets (Fig. 8.2b).
2. The limitations of this protocol include:
   - Limit of detection. Typically, several hundred wells can be counted for each treatment. However, if variants of interest are at a lower frequency in the population ($\leq 10^{-3}$), significantly more wells would be needed, which can be a limiting factor. To solve this issue, plate readers with optical density at 620 nm can be used to automate the counting process, and 384-well plates can be used to increase the number of counts.
   - Turbidity in the wells indicates bacterial growth; however, the absence of turbidity does not necessarily signify bacterial killing. Some phages or compounds may inhibit bacterial growth without causing cell death. Additionally, bacteria may be lysed "from without", even if they are not inherently sensitive to the phage or compound, especially under high phage concentration (Abedon 2011).

3. Any bacterial strain can be used. However, strains prone to clumping should be avoided, as this may hinder single-cell isolation. Any phage targeting the desired populations may be used.
4. As a negative control for phage amplification, add 1 mL of bacteria to a final volume of 100 mL of LB. This serves to ensure the proper bacterial culture growth.
5. Adjust the g-force as needed based on bacterial sensitivity. For fragile bacteria, reduce the centrifugation speed and compensate by increasing the centrifugation time.
6. At least five washes are recommended to reduce the amount of phage in the bacterial culture. However, if no phages are present in the culture, a single washing step is sufficient to remove LB. Note that LB and other media can interfere with cytometer filters by causing auto fluorescence.
7. Some wells may show turbidity after 24 h, but a 48-h incubation is recommended to ensure sufficient growth time for all cells.

Competing Interests P.D.-C. is cofounder of Evolving Therapeutics SL and a member of its scientific advisory board.

Funding PDC was supported by a Ramón y Cajal contract RYC2019–028015-I funded by MCIN/AEI/10.13039/501100011033, ESF Invest in your future, project PID2020-112835RA-I00 funded by MCIN/AEI/10.13039/501100011033, and project SEJIGENT/2021/014 funded by Conselleria d'Innovació, Universitats, Ciència i Societat Digital (Generalitat Valenciana). LMQ was funded by a PhD fellowship from Spanish MCIU FPU19/04611.

References

Abedon ST (2011) Lysis from without. Bacteriophage 1:46–49. https://doi.org/10.4161/bact.1.1.13980

Adler BA, Kazakov AE, Zhong C et al (2021) The genetic basis of phage susceptibility, cross-resistance and host-range in Salmonella. Microbiology 167:001126. https://doi.org/10.1099/mic.0.001126

Azam AH, Tanji Y (2019) Bacteriophage-host arm race: an update on the mechanism of phage resistance in bacteria and revenge of the phage with the perspective for phage therapy. Appl Microbiol Biotechnol 103:2121–2131. https://doi.org/10.1007/s00253-019-09629-x

Bayliss CD (2009) Determinants of phase variation rate and the fitness implications of differing rates for bacterial pathogens and commensals. FEMS Microbiol Rev 33:504–520

Bertozzi Silva J, Storms Z, Sauvageau D (2016) Host receptors for bacteriophage adsorption. FEMS Microbiol Lett 363:fnw002

Bull JJ, Vegge CS, Schmerer M et al (2014) Phenotypic resistance and the dynamics of bacterial escape from phage control. PLoS One 9:e94690

Chan BK, Sistrom M, Wertz JE et al (2016) Phage selection restores antibiotic sensitivity in MDR Pseudomonas aeruginosa. Sci Rep 6:26717. https://doi.org/10.1038/srep26717

Cota I, Sánchez-Romero MA, Hernández SB et al (2015) Epigenetic control of Salmonella enterica O-antigen chain length: a tradeoff between virulence and bacteriophage resistance. PLoS Genet 11:e1005667. https://doi.org/10.1371/journal.pgen.1005667

Davis KM, Isberg RR (2016) Defining heterogeneity within bacterial populations via single cell approaches. Bioessays 38:782–790

Gefen O, Balaban NQ (2009) The importance of being persistent: heterogeneity of bacterial populations under antibiotic stress. FEMS Microbiol Rev 33:704–717

Gordillo Altamirano FL, Barr JJ (2021) Unlocking the next generation of phage therapy: the key is in the receptors. Curr Opin Biotechnol 68:115–123. https://doi.org/10.1016/j.copbio.2020.10.002

Homberger C, Barquist L, Vogel J (2022) Ushering in a new era of single-cell transcriptomics in bacteria. microLife 3:uqac020

Kim M, Ryu S (2012) Spontaneous and transient defence against bacteriophage by phase-variable glucosylation of O-antigen in Salmonella enterica serovar Typhimurium. Mol Microbiol 86:411–425

Kümmerli R, Frank SA (2023) Evolutionary explanations for heterogeneous behavior in clonal bacterial populations. Trends Microbiol 31:665–667

Li N, Zeng Y, Wang M et al (2022) Characterization of phage resistance and their impacts on bacterial fitness in Pseudomonas aeruginosa. Microbiol Spectr 10:e02072–e02022. https://doi.org/10.1128/spectrum.02072-22

Magdanova LA, Golyasnaya NV (2013) Heterogeneity as an adaptive trait of microbial populations. Microbiology 82:1–10

Mangalea MR, Duerkop BA (2020) Fitness trade-offs resulting from bacteriophage resistance potentiate synergistic antibacterial strategies. Infect Immun 88:e00926–e00919. https://doi.org/10.1128/IAI.00926-19

Müller S, Nebe-von-Caron G (2010) Functional single-cell analyses: flow cytometry and cell sorting of microbial populations and communities. FEMS Microbiol Rev 34:554–587

Ruiz LMR, Williams CL, Tamayo R (2020) Enhancing bacterial survival through phenotypic heterogeneity. PLoS Pathog 16:e1008439

Ryall B, Eydallin G, Ferenci T (2012) Culture history and population heterogeneity as determinants of bacterial adaptation: the adaptomics of a single environmental transition. Microbiol Mol Biol Rev 76:597–625

Turkington CJR, Morozov A, Clokie MRJ, Bayliss CD (2019) Phage-resistant phase-variant subpopulations mediate herd immunity against bacteriophage invasion of bacterial meta-populations. Front Microbiol 10:1473

van der Woude MW, Bäumler AJ (2004) Phase and antigenic variation in bacteria. Clin Microbiol Rev 17:581–611

Weigel WA, Dersch P (2018) Phenotypic heterogeneity: a bacterial virulence strategy. Microbes Infect 20:570–577

MIX
Papier aus verantwortungsvollen Quellen
Paper from responsible sources
FSC® C105338

If you have any concerns about our products,
you can contact us on
ProductSafety@springernature.com

In case Publisher is established outside the EU,
the EU authorized representative is:
Springer Nature Customer Service Center GmbH
Europaplatz 3, 69115 Heidelberg, Germany

Printed by Libri Plureos GmbH
in Hamburg, Germany